T0204049

SECOND EDITION

Topical Antimicrobials Testing and Evaluation

SECOND EDITION

Topical Antimicrobials
Testing and Evaluation

Daryl S. Paulson

BIOSCIENCES LABORATORIES, INC.
BOZEMAN, MONTANA, USA

CRC Press
Taylor & Francis Group
Boca Raton   London   New York

CRC Press is an imprint of the
Taylor & Francis Group, an **informa** business

CRC Press
Taylor & Francis Group
6000 Broken Sound Parkway NW, Suite 300
Boca Raton, FL 33487-2742

First issued in paperback 2019

ISBN-13: 978-1-4398-1322-5 (hbk)
ISBN-13: 978-0-367-37787-8 (pbk)

Library of Congress Cataloging-in-Publication Data

Paulson, Daryl S., 1947- , author.
   [Topical antimicrobial testing and evaluation]
   Topical antimicrobials testing and evaluation / Daryl S. Paulson. -- Second edition.
   p. ; cm.
   Preceded by Topical antimicrobial testing and evaluation / Daryl S. Paulson. New York : Marcel Dekker, c1999.
   Includes bibliographical references and index.
   ISBN 978-1-4398-1322-5 (hardcover : alk. paper)
   I. Title.
   [DNLM: 1. Anti-Infective Agents, Local--pharmacology. 2. Drug Evaluation--methods. 3. Statistics as Topic--methods. QV 220]

RM400
615'.7--dc23   2015002093

Visit the Taylor & Francis Web site at
http://www.taylorandfrancis.com

and the CRC Press Web site at
http://www.crcpress.com

Contents

Preface

I began working in the topical antimicrobials area about 33 years ago. This was a time when chlorhexidine gluconate had just entered the market, while iodine was the king. Alcohol products were not used, because they were too irritating and drying to the hand surfaces. PCMX and Triclosan were in much greater demand, because they were extensively used. Also, there were no single applicators for preoperative skin preparations with a wand device for applying them. Instead, the antiseptic came out of a large bottle and was applied with gauze. For surgical scrubs, a brush and nail pick were also used in the application process.

This book, then, has come from a long process of procedures and methods and is an integration of all of them, to inform people of a more comprehensive way to look at developing new products by using the four-quadrant model of subjective and objective domains as well as the individual and whole (society) levels. This can then help one to see the social rules and laws used by the FDA with the cultural beliefs (i.e., things learned in training) of both the personal objective and subjective values. This makes a superior product without making the many costly mistakes others have made.

Also, the important thing for a company to remember is, once it has developed a new product, it must begin working on the next step immediately. I have witnessed many new products developed that negate those currently on the market; for example, a new product caused the surgical scrub brush to become outmoded. The industry moves rapidly in this new direction, and the company fails, because it must compete with many "me, too" products developed soon after.

So this book integrates the fields of microbiology, statistics, regulation and culture and integrates them into the four-quadrant model of human dimension, with sales at the forefront. Do the products work? Are they mild to the skin? Do they fit into the culture? And what are the regulatory requirements concerning these? Finally, how is the product sold?

Acknowledgments

I remember and acknowledge some of the people who were killed in South Vietnam in 1969—incredible people who died far too young. John Burke, John Osterhaus, and Joey Klemenick were U. S. Marines. Semper Fi.

I recognize and thank my parents, Orrin and Sylvia Paulson, who provided much early support for me. The incredible employees at BioScience Laboratories, Inc. of Bozeman and Butte make it all happen, and I am grateful. Tammy Anderson, my assistant and the person who arranged the book, wrote my drafts, and basically put it together, thank you. Finally, to my wife Marsha and our two children, Sasha and Talia, thank you.

About the Author

Daryl S. Paulson, PhD, is a decorated Vietnam combat veteran, a counselor specializing in trauma-associated disorders, and president/CEO of BioScience Laboratories, Inc., with advanced degrees in microbiology, statistics, counseling, human science, and psychology. Dr. Paulson is the author of numerous articles and 14 books, among which are *Biostatistics and Microbiology: A Survival Manual* (Springer, 2009), *Applied Statistical Designs for the Researcher* (Marcel Dekker, 2003), *Handbook of Topical Antimicrobial Testing and Evaluation, Competitive Business, Caring Business: An Integral Perspective for the 21st Century* (Paraview Press, 2002), and *Haunted by Combat: Understanding PSTD in War Veterans* (Greenwood Publishing, 2007). Address correspondence to Daryl S. Paulson, at dpaulson@biosciencelabs.com.

1 General Approach to Developing Topical Antimicrobials

This revised microbiological book addresses four areas of focus needed to produce a quality surgical handwash, a preoperative skin preparation, or a healthcare personnel handwash that meets regulatory criteria for performance, is saleable, and can withstand competition. Because personnel often have been trained in only one of the four areas (Figure 1.1), the other areas are ignored, and a product may fail. This is why we will concentrate on these areas equally.

The business area focuses on how the product can be made profitable in a competitive market. The quality assurance area ensures that the product's manufacture is within tolerance of the specifications and the testing validity for safety and antimicrobial efficacy is efficient. The statistical section evaluates the data for passing or failing the test. Finally, the domain for testing products for their antimicrobial activity resides in microbiology.

Let us now cover these areas in greater detail.

I. BUSINESS

A business strategy provides the direction for the whole product development process. A strategy that would result in product sold at a high price with little or no direct competition is the ideal. The actual process is that the product is sold at a fair price, and it has a strong competitive niche. This strategy will help address questions such as What is the product? How will it be sold? Who will compete against it? and, very importantly, Can it be developed further into second and third generations of products able to maintain a competitive edge?

Probably, the most reliable means of achieving answers is to map product qualities and characteristics (see Figure 1.2).

A diagram such as Figure 1.2 will allow proper categorization of the product respective of its competitors. The product can then be marketed and sold based upon its effective categories. There is no preferred position for any one product to be marketed, as all have strengths and weaknesses. For example, an alcohol product is very fast-acting, but it tends to be skin-irritating and flammable. Other concerns, such as microbial resistance and environmental fate, should be considered as well, such as is the current case with Triclosan. Understanding the strengths and weaknesses of a product will help define a marketing strategy to emphasize the strengths and mitigate the weaknesses.

1

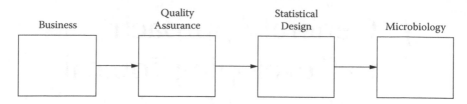

FIGURE 1.1   The four dimensions required in design of a topical antimicrobial product.

	Properties										
	Microbiological Aspects				Business Aspects				Useful Applications		
Product Name	Immediate	Persistent	Residual	Irritation	Cost	Package	Delivery System	Flammability	Surgical Scrub	Preop Prep	HCPHW
Alcohol	×	—	—	×				×			×
Alcohol + Chlorhexidine Gluconate	×	×	×	×				×	×	×	
Alcohol + Iodine	×	×	—	×				×		×	
Chlorhexidine Gluconate	×	×	×						×	×	
Iodine	×	×		×					×	×	
Triclosan	×										×
Parachlorometaxylenol	×			×							×

FIGURE 1.2   Product map. This map is not the most accurate because costs, packaging and delivery are dependent on the product itself.

Figure 1.3, Porter's Five Basic Forces of Competition, diagrams what is happening in any market. The five forces are

1. **Competition, the rivalry among existing firms**. Most professionals believe this to be of utmost importance, yet this view is limited. It must expand to include the other four forces, as will be demonstrated. What we reviewed in Figure 1.3 is vital to competition. For example, if your product is an alcohol or chlorhexidine gluconate (CHG), you might consider marketing a tincture of CHG for a preoperative preparation. This type of product would have the benefit of speed (fast-acting, significant biocidal fungicidal activity) and persistence (long-term effects). It could be marketed for some time before it lost its competitve edge to sell in the current market, including new applications. For example, it could be a great preparation for catheter insertion, and potentially be marketed with a CHG-impregnated bandage to enhance persistence claims.

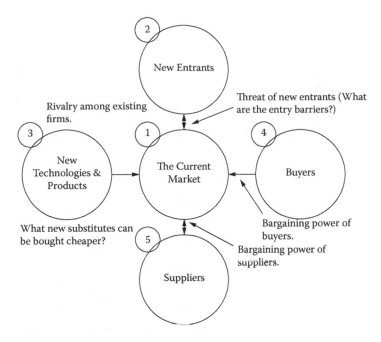

FIGURE 1.3 Porter's Five Basic Forces of Competition.

2. **Potential new entrants**. New competitors evaluate the topical antimicrobial market daily for opportunities. However, many will be deterred by the considerable expense of getting a product, such as a CHG, through the Food and Drug Administration's (FDA) new drug application (NDA) process. This is an example of an entry barrier. Another barrier is the selling and distribution process. How do you market and distribute a product once it is produced?

It is very important to visualize specific performance characteristics of your product. For example, we can make a graph of your product relative to other products in your market (Figure 1.4).

In our example, we have determined that an alcohol + CHG product is the choice. Notice that in this market, only the other alcohol + CHG products can compete against it in all three healthcare applications. But when the same types of products compete within the same market, price is what sells these products.

To sell a product for the most money, one must show it is effective in certain areas not addressed by its competitors. For example, test this product against viruses or methicillin-resistant staphylococcus aureus (MRSA) to show that it is effective, especially if competitors have not focused on these attributes. Or what if your alcohol is not irritating to the skin? Or, what about the product's application for pre-op?

It is important to also look for counterstrategies. If your alcohol-based product causes pain to the user who has breaks in his/her skin surfaces, it will be important for your marketing team to address this concern. Flammability may also be an issue.

	Properties										
	Surgical Scrub	Pre-op Prep	HCPHW	New Developments	Immediate	Persistent	Residual	Flammability	Irritation	Cost of Approval	Cost of Product
Your Product											
Alcohol + CHG	X	X	X		High	High	High	High	High		
Other Products											
Alcohol + CHG	X	X	X		High	High	High	High	High		
Alcohol + Iodine	X	X			High	Med	None	High	High		
Aqueous CHG		X	X	X	Low	High	High	None			
Povidone Iodine		X	X	X	Low	Med	None	None	High		?
Alcohol	X		X		High	None	None	High	High		
PCMX			X		Low	Med	Low	None	Med		
Triclosan			X		Low	Med	Low	None	Med		

FIGURE 1.4   Your product in relation to other products.

3. **New technologies**. The FDA has provided guidelines to determine how topical antimicrobial products intended for healthcare are to be evaluated in the United States, and Health Canada adopted its own set of criteria. Offshore, the European Union has yet a different set of criteria. Currently, Japan is working on developing its own guidelines. As expensive as the processes of meeting these various criteria are, they tend to keep new competitors out of your market. These requirements, when tightly monitored, have prevented many companies from developing new products. This presents a benefit to companies that have spent the money, for it is an effective *barrier to entry*. Remember, if this barrier were not in place, the competition would be even greater in this market.
4. **Buyers**. When there are few buyers for a particular product, they will have immense power. They are able to manipulate a firm by insisting that, unless prices are lowered, they will do their business elsewhere. It is essential that a company constantly expand its market share so as not to find itself in this position.
5. **Suppliers**. It is critical that a firm has multiple sources of raw material and services. If suppliers raise their prices, others with lower prices can be sought.

II. QUALITY ASSURANCE

It is crucial that product testing for safety and efficacy be performed correctly, or the entire study could fail. Successful testing can be assured by (1) inspection of the laboratory and training records of all personnel who perform the work for good laboratory practices (GLP) and good clinical practices (GCP) compliance, (2) ensuring that standard operating procedures (SOPs) are clearly written and implemented, (3) ensuring that the study protocol explains precisely how testing will be performed, and (4) submitting the protocol to the regulatory agency for review and approval prior to conducting the study.

A. INSPECTION OF THE LABORATORY/PERSONNEL PERFORMING THE WORK

When deciding whether to use a commercial laboratory for product testing, it is important that a firm inspect it for appropriate quality assurance standards and practices. The laboratory should demonstrate adherence to GLPs and GCPs. Are appropriate SOPs in place? Documentation of FDA audits should be requested in order to review comments, and consistent adherence by all personnel to internal SOPs must be assured, as well as adequate training for continued improvement. It is also important to ascertain whether the people involved with clinical trials work in teams. Make certain they have worked as a unified team on similar projects in the past.

IRB. Evaluate the institutional review board's (IRB) standard operating procedures and membership composition. Are they appropriate to current regulatory requirements? Does the IRB understand what its responsibilities are? Do the members ever audit an actual study? If subjects should become negatively affected during product testing, does the IRB know exactly what to do and the proper time frame for doing it? Has the IRB been audited recently?

Protocol. Be certain that the protocol provided by the laboratory is in sufficient detail so testing can be performed without ambiguity. Assure that your product's application procedures have been well defined, that statistical models and microbiological procedures are appropriate, and more importantly, that the protocol is consistent with regulatory guidelines. Once you are comfortable with the design specifics, confer with the FDA to have the protocol approved. If this is not done, for whatever reason, the gamble on their acceptance rests solely with you. Usually, the FDA will suggest changes, and once these have been incorporated, the study can commence. A representative of the company should visit the test facility to monitor the critical phase of the study and audit documentation for completeness and accuracy.

Microbiology. Testing of your product will be conducted in the in vitro laboratory and the clinical trials laboratory. It is imperative to understand microbiology because it applies to human vectors of influence. Understanding this is the primary purpose of this book.

Statistics. Before your study is accepted by regulatory agencies, it is important to understand the statistical design used to analyze the data. This is the second general purpose of this book.

III. MOVING QUALITY CONTROL OUTWARD

Some commercial laboratories have internal quality assurance (QA) that is managed by the laboratory but is not part of the in-house QA program. This ensures that each

	Laboratory		QA
	Prestudy	Study	Poststudy QA Review
Quality Processes			X
Quality Processes for Laboratories	X	←X	←X

FIGURE 1.5   Laboratory quality process.

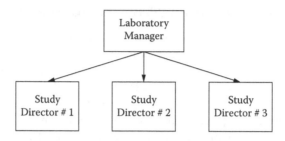

FIGURE 1.6   Study chain of command.

study is done with the highest quality. The quality process provided by the poststudy QA group to laboratory prestudy and study periods looks like Figure 1.5.

Studies must be controlled prior to commencement by moving the poststudy quality control review to a prestudy review of the laboratory. That means nothing for the QA unit, but considerable prestudy work for the laboratories themselves.

The laboratory managers are in charge of all that happens in their laboratories. They delegate control of specific studies to the study directors (Figure 1.6).

The laboratory manager expects the study directors to take control of their studies immediately (prestudy control) and maintain control throughout the study. This will require that the laboratory manager assist the study director in establishing, for example, a Program Evaluation and Review Technique (PERT) system for a study and ensure that it is implemented. As part of the quality planning process, the study director will provide the laboratory manager with personnel requests, and the laboratory manager will schedule accordingly.

Following a study's assignment to a study director, but prior to study launch, the study director, assisted by the associate study director, will assess the prestudy objectives. They will pass responsibility for quality assurance throughout their assigned team, as all are responsible for the quality of the project. This will be done in three ways: (1) quality planning (prestudy), (2) quality control (during the study), and (3) quality improvement (poststudy assessment by means of QA review and a QA-led team meeting).

For example, there are eight prestudy goals that should be addressed for clinical laboratory studies.

1. Protocol
    i. Formal handoff meeting with sales, sponsor, study directors, and management to assure the purpose and scope of the study are understood.
    ii. Appropriate approvals are obtained.

2. Measurement Assurance
   i. All measuring devices are calibrated for current requirements.
   ii. Measuring devices are verified accurate at time of use, as appropriate.
3. Material Assurance
   i. No materials are out of date.
   ii. Supplies are requested and verified as being either on-site or assured to be on-site when scheduled.
   iii. The appropriate microbiological media in requested volumes in specified containers are made.
   iv. All supplies are labeled correctly.
4. Personnel Assurance
   i. People assigned to the study know exactly where to be, when to be there, how long they will be on the job, etc.
   ii. All personnel assigned to a study sign off on the protocol to confirm they understand what they must do.
   iii. Personnel have training documentation relative to the tasks they are assigned.
   iv. Open and honest communication is emphasized at all levels.
   v. If something goes wrong, the personnel know how and to whom to report it.
   vi. Preproject meetings are conducted with staff to discuss the study purpose, tasks, procedures, paperwork, and identification of potential problems, as well as corrective procedures.
5. Environmental Assurance
   i. Make certain you have a clean working area prior to conducting testing.
   ii. Make certain working areas are cleaned prior to and following testing.
6. Method Assurance
   i. Use current FDA, Environmental Protection Agency (EPA), American Society for Testing and Materials (ASTM), and other methods.
   ii. Use the current protocol, any amendments, and applicable internal SOPs.
7. Equipment/Instrument Assurance
   i. Make certain that all equipment (plate counters, incubators, etc.) are validated, functional, and operated appropriately.
   ii. Make certain device calibration procedures are performed correctly and as scheduled.
   iii. Make certain personnel assigned to operation of equipment are trained.
8. Recruitment Assurance
   i. Confirm IRB approval of protocol, informed consent, study description, advertisements, calendars, and any other related documentation.
   ii. Confirm that subject requests are submitted, detailing the number of subjects and backup subjects needed, the study dates, and any instructions specific to a study.
   iii. The study director must meet with the Participant Recruitment Department to review protocol and subject requirements.

A. DURING THE STUDY: CONTROL

The study director and associate study director must remain aware of all processes as the study progresses and be open to input from members of the team. Personnel at any level have authority to inform managers and others of problems and corrections they feel are necessary.

Management goals must be

- Planning
- Organization
- Staffing
- Controlling
- Direction

If goals are achieved, the chances of the study being successful are high.

A good tool to use in finding error is the fishbone checklist (Figure 1.7). It is a cause-and-effect diagram. In its formulation, list the eight prestudy goals as the primary fishbones. Include detail comprising each goal (fishbone).

B. THE FISHBONE CHECKLIST (CAUSE AND EFFECT)

List what is important to achievement of prestudy goals.

The study director checks each of the fishbones to be certain they are adequate before testing commences (Figure 1.7).

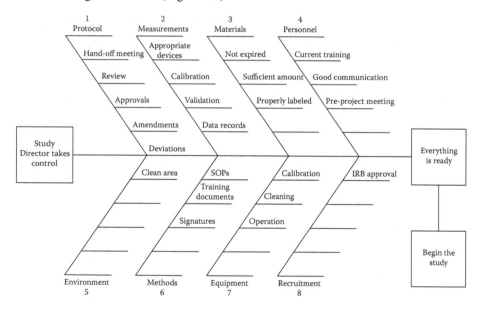

FIGURE 1.7   Fishbone checklist.

C. QUALITY IMPROVEMENT: AFTER THE STUDY HAS BEEN COMPLETED

The study manager will review items for quality before, during, and after the study has been conducted. To this end, a review of the study on the basis of the fishbone checklist is performed. Pareto charts are prepared by QA once a month to track problems, assign them to the aspect of the study in which they occurred, and determine assignable cause and potential solution.

Two types of problems can be observed.

Sporadic problems: These are occasional occurrences with no one assignable reason as to why they occur. Such mistakes will always occur, but they must be kept at low levels.

Chronic problems: These are problems that occur repeatedly. It is critically important that these be discussed frankly with the person or process that creates them. This is the area of retraining.

The types of problems are easily differentiated by imposing the fishbone on a Pareto chart (Figure 1.8). Remember, according to the Pareto chart, about 80% of the errors can be attributed to 20% of the process.

IV. A TIGHTER APPROACH TO DELIVERING A USEFUL PRODUCT

The intense competition present in the topical antimicrobial market, as well as the ever-more demanding FDA performance standards, requires that antimicrobial soap and detergent manufacturers pay very close attention to market requirements [1]. But just what are these additional market requirements? They include factors not often examined, unlike antimicrobial effectiveness, such as low skin irritation, ease of use, whether the product is aesthetically pleasing, and various other subjective values. If such attributes have been addressed, the product likely has been developed with concern and competence, instead of a "me, too" attitude. That is, manufacturers may ignore these critically important factors to produce a product merely to compete with those of their competitors. In the end, this approach often fails; the product is never really accepted into the market [2]. Because the goal is to introduce products that will be highly successful, manufacturers are urged to develop them from a multidimensional perspective.

At least four perspectives should be addressed: social, cultural, scientific, and psychological.

V. THE QUADRANT MODEL

The quadrant model is presented in Figure 1.9.

A. SOCIAL REQUIREMENTS (LOWER RIGHT QUADRANT)

Social requirements, as they apply here, include conforming to the standards of regulating agencies such as the FDA, the Federal Trade Commission (FTC), and

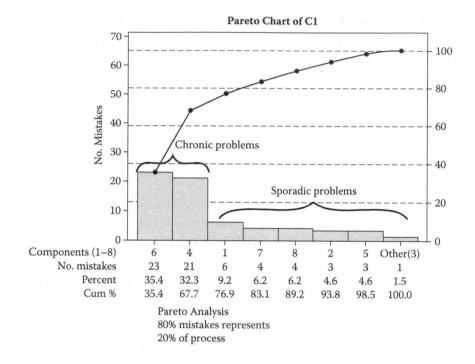

Pareto Chart of C1

Components (1–8)	6	4	1	7	8	2	5	Other(3)
No. mistakes	23	21	6	4	4	3	3	1
Percent	35.4	32.3	9.2	6.2	6.2	4.6	4.6	1.5
Cum %	35.4	67.7	76.9	83.1	89.2	93.8	98.5	100.0

Pareto Analysis
80% mistakes represents
20% of process

Figure Key:

1 = Protocol

2 = Measurement

3 = Material

4 = Personnel

5 = Environmental

6 = Methods (FDA, EPA, ASTM)

7 = Equipment

8 = Subject Recruitment

FIGURE 1.8 Pareto chart.

the Environmental Protection Agency (EPA), as well as the rules, laws, and regulations they enforce (Figure 1.9, lower right quadrant). Before completing product design, it is critical to understand the regulations governing the product, its components, and their concentrations, as well as determine product stability and toxicological properties. For example, a new drug application (NDA) is required in order to market a regulated drug product. For over-the-counter (OTC) products, the active drug must be allowable and its levels within allowable limits. Additionally, the FDA's Tentative Final Monograph (TFM) for OTC products or the "to-be-determined" recommendations of the Cosmetic Toiletries and Fragrance Association (CTFA) and the Soap and Detergent Association (SDA) must be addressed.

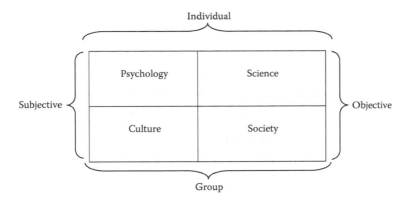

FIGURE 1.9 The quadrant model.

B. CULTURAL REQUIREMENTS

Cultural requirements are very important. It is the culture that buys the product—but it is often ignored (Figure 1.9, lower left quadrant). Cultural and subcultural requirements include shared values, beliefs, goals, and the world views of a society or subgroup of society [3]. Shared values such as perceived antimicrobial effectiveness have much influence on consumers [4]. These values are generally of two types: manifest and latent [5, 6]. *Manifest* (surface) values are conscious to the consumer. For example, a consumer buys an antimicrobial soap to be *cleaner* than they perceive they can be using a nonantimicrobial soap. But deeper and more fundamental values lie behind those that are manifest. These are referred to as *latent* values and generally are unconscious to consumers. In this case, *cleaner* may encompass such basic needs as being accepted, loved, and worthwhile as a person, spouse, or colleague.

Most manifest and latent values we share as a culture are magnified by manufacturers' advertising campaigns in which consumers are motivated in terms of shared manifest and latent values. For example, if homemakers perceive that they are taking better care of the children by having them use antimicrobial soaps (a manifest value), and if they feel more valued, more loveable, or more needed by their family (latent values), they will be motivated to purchase the product. Finally, much of what consumers believe to be true is not grounded in objective reality [7]. Most of these beliefs are formed from their interpretation of mass media reports, opinions of others, and explanations of phenomena from various sources [8, 9].

C. SCIENCE

These include the physical aspects of a product—its ease of use, some aesthetic characteristics, antimicrobial efficacy, some irritation potential or moisturizing ability, in-use effects (e.g., staining clothing, gowns, and bedding), and so on (Figure 1.9, upper right quadrant). It is important that products be designed with the individual end user in mind [2]. Hence, products must be easy to open (if in a container), easy to use, and effective in their intended use (e.g., by food servers, home consumers, medical personnel, and surgical personnel).

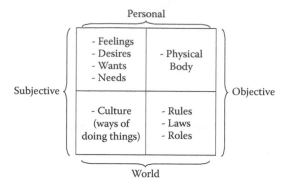

FIGURE 1.10   Quadrant model of attribute categories.

D. PSYCHOLOGY

This category includes one's personal interpretation of cultural and subcultural world views (Figure 1.9, upper left quadrant). Relative to antimicrobials, these include, for example, subjective likes and dislikes of characteristics such as the fragrance and feel of the product, the perceived quality of the product, and other aesthetic considerations [4]. As with cultural attributes, manifest and latent values are operative in this category. Hence, if one likes the springtime fragrance of a consumer bodywash product (manifest), the latent or deeper value may be that it makes one feel younger, and therefore more physically attractive and desirable as a person.

These four attribute categories can be presented in a simple quadrant form (Figure 1.10). Each quadrant interacts with the other three quadrants. For example, cultural values influence personal values, and vice versa. Cultural and personal values influence behavior, and behavior influences values.

All four of these quadrants must be integrated into one's perspective. The problem is that one usually does not learn the synthesized perspective in their training, so they learn to reduce the world view to view one quadrant and ignore the others, according to their training.

VI.   A RECURRING PROBLEM: REDUCTIONISM

A recurring design problem happens during the product's development because the details become compartmentalized, because our society—we—have learned to specialize, not integrate. For example, reductionism occurs when personnel such as engineers, chemists, microbiologists, statisticians, and lawyers focus solely on the objective parameters (the two right quadrants), and therefore interpret subjective parameters in objective terms. That is, they reduce the subjective domains to objective ones [5, 6, 10]. Equally problematic are marketing personnel, industrial psychologists, and social scientists who focus their efforts in the subjective domains and interpret objective domains in subjective terms. Reductionism is often rooted in subcultural bias. So those who were trained in the physical sciences tend to have incorporated an objective world view, while those trained in social sciences usually will have a subjective one. But notice that both world views are true. Both must be recognized and integrated into the product's development. This, in no way, suggests that each specialist must work in all four quadrant domains; that is unrealistic. Each should focus on his/her own domain (e.g., chemists in the two right quadrants, and marketing in the two left quadrants). The person or group in charge of the product's development, however, must be conscious of the four domains and integrate them into the design of the product. With this in mind, let us examine how the quadrant model may be of assistance in the design of topical antimicrobial products. This should be done if you want your product to be truly successful; you must consider and integrate these four quadrants.

A. PRODUCT CATEGORIES

There are three general categories of topical antimicrobial products recognized by the FDA:

1. preoperative skin-prepping formulations;
2. surgical scrub and antiseptic formulations; and
3. healthcare personnel handwash and hand antiseptic formulations.

The CTFA/SDA groups have introduced what they term the *Healthcare Continuum Model*, which adds three product categories to those listed above:

1. antimicrobial bodywashes;
2. general-use antimicrobial hand soaps; and
3. food-handler hand soaps.

It remains a question whether these three additional categories will be formally recognized by the FDA, but because they are anchored in personal and cultural values, they are important. It is fair to say that most Americans want to feel clean and be clean. This, to many, means using antimicrobial body soap for showering and bathing, as well as for hand cleansing.

Moreover, there is considerable pressure on governmental agencies to make the food we eat safer. Consumers remember clearly the deadly problems created by ingesting *Escherichia coli,* Serotype O157/H7, in contaminated hamburger [11]. They want governmental protection to prevent such food-borne epidemics from recurring [12, 13].

These areas will now be expanded into the four quadrants.

1. Preoperative Skin-Prepping Formulations

a. Personal Objective (Figure 1.10 Upper Right Quadrant)

Preoperative skin preparation formulations are designed to degerm an intended surgical site rapidly, as well as provide a high level of persistent antimicrobial activity—up to 6 hours—post-skin prepping [14]. In terms of the quadrant model presented earlier (Figure 1.10), these requirements belong to the personal objective region (upper right quadrant).

The preoperative skin preparation product should be convenient and easy to use for surgical staff consumers. Given two products with equal antimicrobial efficacy and the same cost, the one that is easier to use (e.g., has a shorter skin-prep time, can be more easily seen on the skin due to a coloring agent, or is a one-step procedure) will be preferred over the other.

An often overlooked aspect of product acceptance is customer service. It is ironic that many firms spend large sums of money to assure a product meets its intended purpose but fail to meet a customer's needs. Orders are delayed or lost, return phone calls are promised but not made, and so on. In short, customers are ignored.

b. Social Requirements (Figure 1.10 Lower Right Quadrant)

The documentation requirements for getting a topical antimicrobial product approved to market in this country are fairly straightforward. For example, the product must be manufactured under good manufacturing practices (GMPs), and laboratory testing of the product must be performed under GLPs. If the product is considered a drug, an NDA or an abbreviated new drug application (ANDA) must be filed. Regulatory agencies such as the FDA also require that a battery of specific tests be conducted. These include such things as clinical trials, in vitro time kill kinetic studies, and minimum inhibitory concentration (MIC) studies versus bacteria and fungi.

c. Cultural Requirements (Figure 1.10 Lower Left Quadrant)

The products must be perceived as being of high quality, a perception based squarely in the world view of the culture, including shared values, beliefs, and goals, as well as those of any subcultures, where applicable [3]. Often, however, cultural and subcultural world views are not easily perceived by those belonging to that culture or subculture.

Taking an example relevant to topical antimicrobial product development, if the product ingredients are viewed as *artificial*, when *natural* ingredients are more valued, acceptance of the product containing artificial ingredients is likely to be impaired. Also, social institutions within a culture—family, religious group, social group, and political group—are important, and manufacturers should take care not to alienate them with *thoughtless* advertising. Shared beliefs and values concerning the firm that makes the product are critical. If a manufacturer is perceived by consumers as a quality, caring firm, it will be easier for it to introduce and sell new products than if it is perceived otherwise.

Much of social reality is constructed—made up—by cultures [8]. The degree to which reality can be a cultural construct, however, is bound by objective reality [5, 6]. What we believe to be true or real, therefore, often is only partially true or real. From a sociological perspective, there are three stages in socially constructed reality: externalization, objectivation, and internalization [3]. Externalization is the initial stage in which a theory or opinion is accepted *tentatively* as being true. In the next stage, *objectivation*, the theory or opinion is accepted as "fact." And, finally, *internalization* is when the "accepted facts" are incorporated into a person's psyche as "absolute" truth.

It is often easier to use socially constructed reality as an ally to marketing programs than to educate people concerning truth. For example, with household antimicrobial products, when consumers read an advertisement stating the product "kills 99.9% of all germs," they literally believe only a very small number of disease-causing microorganisms survive. Yet, in truth, an error has been committed—that of applying a linear *proportion* measurement (% reduction in microorganisms) to an exponential distribution. The method of calculating the percent reduction uses a percent (linear) on an exponential distribution (nonlinear), often reexpressed in log_{10} scale. If a linear percent measurement is calculated on a *linearized* exponential distribution, the common interpretation would be correct, but the percent reduction would be far less than 99.9% [13]. For purposes of advertising, the illusory presentation obviously sounds better to consumers who will buy according to their constructed reality.

d. Personal Subjective (Figure 1.10 Upper Left Quadrant)

Individual members of a group (e.g., surgical staff) need to find value and construct positive beliefs about a specific product if it is to be successful. Therefore, their perspective is important. Individuals must believe that the product was designed with them in mind, and providing them specific examples, comparisons, and test conclusions is a very effective way of externalizing, objectifying, and finally, causing internalization of the advertising. But one must be certain that the claims can be supported and are grounded in objective reality [2, 16, 17]. Recall that socially constructed reality is bounded in objective reality [6]. Hence, when people are told that a product is easy to use, they must find this to be so when they use the product, or their beliefs will shift to match the reality of their experience.

2. Surgical Scrub Formulations and Hand Antiseptics

a. Personal Objective (Figure 1.10 Upper Right Quadrant)

Surgical scrub formulations are designed to remove both transient and normal (resident) microorganisms from the skin of hands and lower arms [14]. Surgical scrubs, to be effective, must demonstrate immediate, persistent, and residual antimicrobial properties, and must be low in skin irritation effects when used repeatedly over a prolonged period. (Note: The term *scrub*, as used in the following discussion, includes leave-on hand sanitizers.)

A product's immediate antimicrobial efficacy is a quantitative measurement of both the mechanical removal and immediate chemical inactivation of microorganisms residing on the skin surface. The persistent antimicrobial effectiveness is a quantitative measurement of the product's ability to prevent microbial recolonization of the skin surfaces, either by microbial inhibition or lethality. The residual efficacy is a measurement of the product's cumulative antimicrobial properties after it has been used repeatedly over time. That is, as the antimicrobial product is used over time, it is adsorbed onto the stratum corneum of the skin and, as a result, prevents microbial recolonization of the skin surfaces.

b. Social Requirements (Figure 1.10 Lower Right Quadrant)

The documentation and legal requirements for surgical scrub products are similar to those for the preoperative skin preparations. In general, products must meet the efficacy requirements in clinical trials utilizing healthy human test subjects, as well as the same series of in vitro tests. Like those for preoperative skin-prepping formulations and healthcare personnel handwash formulations (to be discussed next), the actual requirements are presented in the FDA's Tentative Final Monograph for OTC Products (21 CFR, Parts 333 and 339).

c.   Cultural Requirements (Figure 1.10 Lower Left Quadrant)

In general, these are the same as those presented in the discussion of cultural aspects relating to preoperative skin-prepping formulations.

d.   Personal Subjective (Figure 1.10 Upper Left Quadrant)

These are the same as issues covered in the preoperative skin preparation portion of this book. Because surgical staff members actually use the product on themselves, however, its aesthetic attributes will be more important than for preoperative skin-prepping formulations.

It is important that relevant but subjective product attributes be understood on the basis of evaluations by the potential product users. This will require subjective testing to evaluate the sensory attributes of the product—its container, its packaging, and other aesthetic concerns—in order to engineer a more desirable product [4]. Some of the variables of interest are presented in Table 1.1.

Developing a list of attributes from which panelists may select is difficult. One effective way of doing so is through the use of focus groups [4], which consist of a small number of surgical staff members (5 to 10) literally sitting down together and coming up with attributes of a surgical scrub product that are important to them.

TABLE 1.1
Evaluation Criteria

Sensory Evaluation	Acceptance Attributes
1. Clear	1. Like appearance
2. Opaque	2. Like fragrance
3. Strong smell	3. Like texture
4. Light smell	4. Like feel after wash
5. No smell	5. Like overall
6. Oily feel	6. Purchase intent
7. Dry feel	
8. Soft feel	
9. Lathers well	

Performance (nonantimicrobial)	Image
1. Lathers well	1. Effective
2. Feels good on hands	2. Unique
3. Does not irritate hands	3. Good for hands
4. Conditions hands	4. Won't ruin gloves
5. Removes oil from hands	5. Won't irritate hands
6. Feels clean	6. High-quality product
7. Rinses easily	

Once the subjective characteristics deemed important for success of the surgical scrub product have been determined, it is important to actually perform preference testing by enrolling surgical staff personnel in a study to evaluate several configurations of a manufacturer's surgical scrub product, as well as those of the competition.

3. Healthcare Personnel Handwash Formulations and Hand Antiseptics

a. Personal Objective (Figure 1.10 Upper Right Quadrant)

Healthcare personnel handwash formulations and leave-on antiseptics for use without water are intended to quickly eliminate any transient, pathogenic microorganisms picked up on the hands of a healthcare provider from patient A to prevent their passage to patient B or environmental surfaces. Hence, the product is intended to break the disease cycle at the level of the contaminated healthcare worker's hands by removing potentially infectious microorganisms.

The product must demonstrate low skin irritation upon repeated and prolonged use (20 to 30 washes per day for 5 consecutive days). Mildness to the hands, however, is often attained at the price of reduced antimicrobial effectiveness. Hence, an optimal formulation, which is the objective of product development, provides significant reductions in contaminative microorganisms and is gentle to the hands.

Mildness can generally be built into a handwash or hand antiseptic in three ways [18]:

1. By proportionally reducing irritating active ingredients such as CHG, iodophors, or alcohol. For example, instead of 4% CHG, customarily found in surgical scrub formulations, a 1% or 2% CHG formulation may be developed.
2. By adding skin conditioners or emollients. These tend to counteract the irritating effect of antimicrobially active compounds, resulting in a product that is more gentle and mild to skin.
3. By using a combination of these two methods (i.e., a reduction in the levels of active ingredient and the addition of emollients and skin conditioners. Because so many healthcare personnel products were originally marketed as (much harsher) surgical scrubs, the market is vulnerable to manufacturers or customers who will develop or select products appropriately designed for their intended application.

b. Social Requirements (Figure 1.10 Lower Right Quadrant)

The documentation and regulations required to market a healthcare personnel handwash product are covered in the discussion of social requirements for preoperative-prepping products.

c. Cultural Requirements (Figure 1.10 Lower Left Quadrant)

The healthcare personnel handwash or hand antiseptic must be perceived as a highly effective antimicrobial formulation, capable of removing and/or killing the microorganisms with which healthcare personnel may become contaminated [18]. Important shared beliefs/values relevant to these formulations result from perception of their ability to inactivate certain pathogenic microorganism species and are believed to be the important indices of product effectiveness. The perception of importance for many of these is not grounded in objective fact, but, nevertheless, they remain important because they are *believed* to be important. For example, the anaerobic bacterial species *Clostridium difficile* will not grow in the presence of free atmospheric oxygen. However, the species is capable of producing a very resistant spore, and healthcare personnel handwash formulations and hand antiseptics, as a rule, are not sporicidal. Many healthcare workers believe that such formulations should be able to kill *C. difficile* spores if approved for use by healthcare personnel.

Another aspect perceived as valuable in healthcare personnel products is mildness to the hands. The product must not irritate the users' hands or create a general impression of harshness. Most healthcare personnel—particularly physicians—are very conscious of their hand-skin integrity. They do not and will not use products they perceive as harsh. Some of the important attributes that must be determined relating to product quality include those presented in Table 1.1; hence, a focus group of 5 to 10 healthcare personnel should meet in focused forums to determine qualities of importance to be built into a product.

d.   Personal Subjective (Figure 1.10 Upper Left Quadrant)

Once general subjective characteristics of importance to healthcare personnel have been determined, it is important that one actually perform preference testing [4]. The goal is to engineer a product that healthcare personnel prefer over those of the competition. This is done by enrolling a number (the greater, the better) of healthcare personnel as panelists to provide subjective evaluation of several configurations of a test product and, preferably, even those of competing products. Other important considerations regarding personal subjective attributes have been discussed in the section relating to preoperative skin-prepping formulations.

4.   Food-Handler Handwash Formulations and Hand Sanitizers

a.   Personal Objective (Figure 1.10 Upper Right Quadrant)

The potential for food handlers to be vectors in the transmission of food-borne disease is very significant [19, 20]. Contaminating microorganisms are responsible for outbreaks of infectious diseases passed from food handlers to consumers via the food they eat. One of the most common sources of this is food handlers contaminated with enteric microorganisms from hand contact with their own feces.

A significant problem in the food-handling arena is that many who handle food do not cleanse their hands adequately after defecation or urination. To compensate for this, many food establishments have required that food-handling personnel wear barrier gloves [19]. A vinyl or latex barrier glove that is intact (has no holes, rips, or punctures) and uncontaminated will undisputedly provide better protection from microbe transmission. But vinyl food-grade gloves, those most frequently used in the food service industry, commonly have preexisting pinhole punctures that compromise the barrier protection. Additionally, these gloves are easily ripped, torn, or punctured as personnel perform their normal duties and, in many cases, such damage remains unknown to the wearer. Exposure to heat also has been reported to alter the integrity of barrier gloves significantly, making them brittle and thus prone to breakage. Hence, the actual protection provided by barrier gloves is often much less than assumed. For example, in a study conducted at this facility, the hands of volunteer human subjects were inoculated with a strain of *Escherichia coli* [19]. The subjects then donned vinyl food-handler gloves, each of which had four small needle punctures. Within 5 minutes, the outside of the gloves were sampled for microbial contamination. The results of this testing demonstrated that significant numbers of *E. coli* can be transferred from contaminated hands onto the outer surfaces of the gloves, if even small holes exist in the gloves.

On the other hand, wearing gloves actually may serve to increase the potential for disease transmission. As one wears vinyl or latex gloves for an hour or so, the microorganism populations on the hands within increase dramatically because the gloves prevent aeration of the hands, thereby increasing the levels of moisture, nutrients, and various other factors necessary for microbial growth. This phenomenon has long been recognized in the medical field, where mandatory handwashes using antimicrobial soap are required prior to gloving. Logically, as population numbers of both resident and contaminating microorganisms increase, so does the potential for disease transmission. Hence, relying solely on barrier gloves to prevent disease is not prudent.

Hand cleansing has been used for years to prevent food-borne illness [12, 20]. (Note: The term *hand cleansing*, as used in the following discussion, includes application of leave-on hand sanitizers.) Effectiveness is dependent on two factors: (1) the physical removal of microorganisms and (2) the immediate inactivation of microorganisms through contact with an antimicrobial ingredient in the product. But that is not all; many antimicrobial chemistries continue to prevent transient microbial recolonization of the hand surfaces after hand cleansing by either microbial inhibition or lethality. If a non-antimicrobial soap product is used, for instance, only the mechanical removal of microorganisms is significant.

In general, hand cleansing is very effective in removing contaminating microorganisms, given the application is performed *correctly*.

The antimicrobial activity of food-handler products should be very high. This is because food handlers tend to perform hand cleansing less frequently and less thoroughly than do healthcare personnel, often leaving dirt and grime around and under the fingernails, which then may serve as a contamination source.

b. Social Requirements (Figure 1.10 Lower Right Quadrant)

Currently, regulation of soap products created for use in the food industry is somewhat vague. The U.S. Department of Agriculture (USDA) "E" rating system is obsolete, as the regulatory function has passed from the USDA to the FDA. Currently, no study design has been accepted by the FDA for use of human subjects to demonstrate efficacy of food-handler handwash or sanitizer products. Many manufacturers have adopted a modification of the healthcare personal handwash using *Escherichia coli* as the contaminating microorganism species, instead of *Serratia marcescens*, the contaminating microorganism most commonly used in those evaluations. And the toxicological properties of the product ingredients are very important because residues may be transferred from the hands of food handlers to the food they handle.

c. Cultural Requirements (Figure 1.10 Lower Left Quadrant)

In general, these are similar to those discussed relative to healthcare personnel hand-washes and leave-on antiseptics; a major difference, however, is the high degree to which medical personnel value clean hands. Food handlers often have no real shared values relating to clean hands. Hand cleansing is just something they must do while on the job. This is probably because food-handler positions generally are not filled by professionals, but by individuals from lower socioeconomic levels, young people, and often, unmotivated individuals [19].

To a large degree, culturally shared values of food handlers will have to be instilled by the food industry itself. This process will include training (upper right quadrant aspect), which will stimulate a value in doing it right (upper left quadrant aspect). In addition, in order to reduce the potential for disease transmission from fecal contamination, the following four steps will be useful in creating shared values/beliefs for these workers and the industry [12, 19].

1. Both gloving and hand cleansing with an effective antimicrobial product should be required for those performing high-risk tasks such as handling, cooking, or wrapping food. While neither practice is fail-safe, it is probable that the combination will provide more protection against disease transmission than either used alone.

   As mentioned earlier, observations in testing performed in this laboratory showed that population numbers of contaminative *E. coli* actually increased on the gloved hands when glove changes were performed at 1- or 3-hour intervals. However, concurrent testing showed that a thorough hand-wash using an effective antimicrobial product prior to gloving prevented significant growth of the contaminative microbes on the hand surfaces over the course of 3 consecutive hours of wear. The obvious conclusion is that, before gloves are put on, a thorough hand cleansing should be performed using an effective antimicrobial soap. Even so, when feasible, no direct hand or glove contact with food should occur, and sanitized serving tongs or other utensils should be used for its manipulation.

2. Mandatory, ongoing sanitation training and education should be required of all employees. This is particularly necessary with inexperienced or unmotivated workers. Emphatically, without the active participation of employees, achieving adequate sanitation standards will be very difficult.

3. A high degree of personal hygiene should be required of food service personnel. Uniforms should be clean and changed often, employees should bathe or shower often, and they should not perform high-risk tasks when they are ill. High-risk tasks include hand/glove contact with food.

4. A quality control program supervised by qualified personnel should be initiated at each food service facility to monitor hand cleansing and gloving practices. Written SOPs should be drafted, and all employees formally trained in hand cleansing and gloving procedures. Training should be documented in employees' training records, and hand cleansing procedures posted in conspicuous places, particularly near sinks used for handwashing.

d.   Personal Subjective (Figure 1.10 Upper Left Quadrant)

Preference testing will be extremely valuable to determine what types of product attributes food handlers prefer [13, 18]. The goal is twofold: (1) to build a product that food handlers will want to use and (2) to build a product that is preferred over those of the competitors. Because relatively little is known about the subjective preferences of food handlers, it is important that such testing include enrolling a reasonably large number of food handlers in a study to evaluate various products sequentially. For example, begin with fragrance preferences. When one or two preferred fragrances are identified, add lathering or ease-of-use characteristics to the evaluation. Then, with these attributes identified, evaluate the feel of the hands after washing. This type of interactive process can be conducted for all other attributes deemed important.

5.   Antimicrobial Handwash and Bodywash Formulations

a.   Personal Objective (Figure 1.10 Upper Right Quadrant)

Whether these categories become officially acknowledged by the FDA will likely not matter, because consumers want antimicrobial hand cleansers and body soaps. The antimicrobial compounds generally used in these types of products are parachlorometaxylenol (PCMX), triclosan, and isopropyl alcohol, all of which have been used in hand-cleansing products for many years. The antimicrobial hand cleansers and body soaps must also be of low irritation potential.

Perhaps the biggest problem with antimicrobial hand cleansers is that various claims made by some manufacturers are misleading and even false. It is critical that manufacturers take responsibility for manufacturing products that perform well in clinical trials both for efficacy and irritation potential.

b.   Social Requirements (Figure 1.10 Lower Right Quadrant)

Because FDA requirements do not exist for antimicrobial hand soaps and sanitizers, regulation of quality is left to the industry, particularly the Personal Care Products Council (PCPC) and the SDA. These organizations have truly taken responsibility for setting realistic standards to which members voluntarily conform. However, there continues to be concern with various manufacturers unable to assure and verify their label and advertising claims to the satisfaction of the Federal Trade Commission (FTC).

c.   Cultural Requirements (Figure 1.10 Lower Left Quadrant)

Consumers have an obvious, pervasive need to feel clean. The shared values, due mainly to effective advertising campaigns, have convinced people that using antimicrobial hand/body cleansers will make them cleaner than if they do not. And, as a bonus, if they use antimicrobial bodywashes, they will not offend others with body odor.

There is nothing wrong with cultural beliefs that antimicrobial cleansers are better, as long as manufacturers truly strive to live up to their stated labeling/advertising claims. That is, they must support their claims with data collected from statistically valid studies using human subjects, as well as from in vitro testing for microorganism sensitivity to the product.

d.   Personal Subjective (Figure 1.10 Upper Left Quadrant)

Soap and leave-on sanitizer manufacturers have mounted a very strong campaign in support of antimicrobial products. And frequently, manufacturers of these consumer products have taken the time to understand the personal subjective and cultural values, using that knowledge to launch highly successful marketing campaigns.

VII.   CONCLUSION

In developing topical antimicrobial wash products, it is important that a complete, multidimensional approach be taken [9]. This will help ensure that the resulting products are designed for the specific needs of the market and that those needs are met. In this way, the product is more likely to have a long, useful, and profitable life.

2 Skin Properties

The skin on which the microorganisms are found is in a constant relationship with the effects of antimicrobial products.

Figure 2.1, simplistically, depicts the relationship as a triangle. The skin and the antimicrobial come together to form a unified process for killing or ridding skin surfaces of infectious microorganisms. When continuity of the skin is breached, by surgery for example, microorganisms can gain access to tissues and the bloodstream. Prior use of a topical antimicrobial reduces this risk.

In this chapter, we will discuss the importance of the anatomical structure of the skin, or integument, as the first leg of the triangle. Human skin, the largest single organ, provides a continuous covering over the body. The skin constitutes about 16% of the total body weight, having an area for average-size adults of approximately 1.8 m^2 [21].

The skin has several important functions. It presents a barrier protecting an organism's interior from bacterial, viral, and fungal invasion, and it plays a critical role in maintaining thermoregulation and water balance. Finally, the skin is a fragile sense organ.

Because skin is under slight tension, it is observed to shrink when a section is removed from the body. The surface structure of skin appears to contain innumerable etch lines and wrinkles, which are derived from underlying fibers in the tissues. In most skin areas, the fibers are irregular such that etching forms a distinct *mosaic* pattern. However, on surfaces such as the finger and toe pads, and to a lesser extent the palms and soles, is a unique regular and consistent arrangement of skin ridges and furrows acquired before birth and retained throughout life.

More obvious skin etch lines and folds are seen at what are termed *skin joints*. These are areas where the skin is bound firmly to the deeper tissue layers. The skin at other parts of the body is loose, separated from muscles and organs by a layer of subcutaneous connective tissue, the hypodermis, which is interspersed with varying amounts of fatty tissue. In anatomical locations where the skin is relatively loose, it also tends to be elastic. This elasticity is gradually lost with age, however, and the skin becomes thin, loose, and folded.

I. STRUCTURE

The skin consists of two general layers, the epidermis and the dermis, which are composed of cells in varying stages of keratinization [22]. Both layers contain elastic fibers which, in large part, determine the elasticity and firmness of the skin. The epidermis contains no blood or lymph vessels. However, the inner layers consist of metabolically active cells, strongly bound together by spot desmosome junctions. Underlying the epidermis is the dermis, which is made up of connective tissue

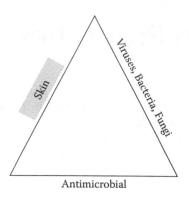

FIGURE 2.1   Skin–product–microorganism relationship.

containing a rich supply of both blood and lymphatic vessels and varies in thickness in different regions of the body. Human dermal skin comprises two layers with rather indistinct boundaries, the outer papillary layer and the deeper reticular layer. The papillary layer is thin and composed mainly of loose, connective tissue. This layer contains fibroblasts, mast cells, and macrophages, as well as other migrant leukocytes. This, by the way, is the housing of the reticuloendothelial system, which is the network of the immune system. The reticular layer is thicker than the papillary layer and is composed of dense, irregularly arranged connective tissue. It contains more collagen fibers and fewer cells than the papillary layer.

The subcutaneous tissues lying beneath the dermis vary in thickness, as well as histological makeup, in different anatomical regions and contain a vascular network that supplies nutritive, immunological, and coagulative properties. Additionally, a nerve network connected with various subcutaneous fibers brings sensory stimuli from the skin surface to the brain and conveys certain autonomic motor stimuli from the brain to various specialized involuntary muscle structures. The interface between the epidermis and dermis is highly irregular.

Let us now look at the epidermis and dermis in greater detail (Figure 2.2).

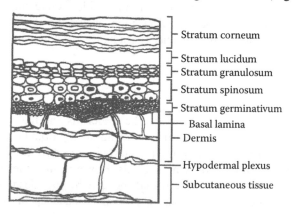

FIGURE 2.2   Cross section of the composition of human skin.

II. EPIDERMIS

The epidermis is composed of a stratified epithelium made up of multiple layers of cells called *keratinocytes*. These cells are continually renewed by cell mitosis, or cell division, in the basal layer. The outer layer of the skin (epidermis) has a thickness varying from 75 to 150 μm. It normally presents an impenetrable barrier to micro-organisms, while also providing a suitable environment for a variety of *resident* microorganisms. These residents serve to prevent epidermal colonization by other pathogenic microorganisms by their rapid colonization of the skin's surface [20, 21].

The several layers of the epidermis can be visually distinguished from each other. The innermost layers consist of metabolically active cells strongly bonded by spot desmosome junctions. As they approach the skin's surface, they die, and their cell bodies are continually shed. Their transit time from the base of the epithelium to the surface averages 20 to 30 days. The stratified epidermis lies on top of the dermis (Figure 2.2), from which it is separated by the basement membrane, or basal lamina, a continuous and structureless sheet distinct from the cytoplasmic membranes of the adjacent cuboidal or columnar cells. The epidermis is supported by numerous extracellular fibrils, which can be regarded as part of the basal lamina, as well as by some of the uppermost dermal epithelial cells [22, 23].

The basal lamina supports a layer of basal cells (cuboidal or low columnar), the stratum basal of the epidermis (Figure 2.2). These cells appear to lack continuous attachment to one another, but are connected by many fingerlike projections from the upper epidermal layers, and contain many dense particles consisting of ribo-nucleoprotein. The cells are metabolically active, but histologically undifferenti-ated at the deepest level. The stratum basal is seen to consist of one, and in some cases several, cell layers in advancing stages of cellular differentiation as the cells move toward the skin surface to constitute the stratum spinosum. The stratum spinosum consists of irregular-shaped cytokeratine cells (relatively large cells con-taining large numbers of keratin filaments) that radiate and become increasingly elongated and flattened as they move outward toward the surface. These cells are firmly bound together by cytoplasmic expansions filled with bundles of filaments called tonofibrils. Desmosomes cover the cell surfaces, giving the cells a prickly, studded appearance in the light microscope. The progressive cellular differentia-tion is accompanied by the production of low-density intercellular "cement," which augments the cellular adhesive properties, assuring that a firm, continuous, tough layer is formed. Intercellular bridging structures are also present and assist in the formation of a compact tissue.

Superior to the stratum spinosum (three to five bundles) lies the stratum granulosum [22]. This layer consists of one to several layers of granular cells usually morpho-metrically larger than cells found in the stratum spinosum. The cells are nucleated, and the cytoplasms contain numerous aggregates of poorly characterized material called keratohyalin, as well as tonofibrils that appear fragmented, both of which play a role in the process of skin keratinization. The granular cells of the stratum granu-losum are in the penultimate stages of keratinization.

The four to six flat layers of the epidermis residing above the stratum granulo-sum consist of dying and dead cells in various stages of keratinization. The deepest of these keratinized layers, the stratum lucidum (not shown in Figure 2.2), is present only in areas of thickened skin, such as the palms and soles of the feet. The cells of the stratum lucidum appear translucent and extremely flattened, containing only disintegrating cell nuclei, and keratohyalin material not visually distinguishable with a light microscope [22].

The stratum corneum, the most external component of the epidermis, consists of many layers of flattened, dead squamous cells containing no nuclei or cytoplasmic organelles. These cells contain large amounts of keratin and are firmly attached to one another. The intercellular spaces are filled with multiple lipid bilayers, packed as an alternating pattern of nonpolar hydrocarbon regions and polar head group regions. The lipids involved are ceramides, cholesterol, cholesterol sulfates, and free fatty acids organized into multilaminated shells. The bilayers consist of straight, loosely packed, saturated hydrocarbon chains, without structural perturbation among the hydrophobic chains. This highly ordered, rigid structure results in a unique impermeability to many compounds, including water. In some cells, remnants of nuclear membranes can be observed with a light microscope. The most external cells of the stratum corneum are constantly shed as minute skin particles (squames) and, in certain conditions, as visible flakes and sheets (e.g., dandruff). This cell loss, or exfoliation, is compensated by constant replacement of cells from the epidermal layers beneath. That is, epidermal cells continually move toward the skin surface in successive stages of cellular differentiation and finally death, until they are exfoliated to the environment. This exfoliative process provides a constant replacement of nutrients for resident microorganisms (e.g., *Staphylococcus epidermidis*) colonizing the skin surface. For the most part, the available skin surface nutrients are insufficient to support growth of fastidious, nonresident microorganisms [25].

The epidermis varies in thickness dependent on anatomical region and gender. It is thicker on the dorsal and extensor aspects than on the ventral and flexor aspects of the body and, as a rule, is thicker in males than in females. It also changes in thickness with age; an infant's skin is thinner than an adult's, but as an individual ages, the skin thickness declines, and it becomes increasingly fragile [22].

The regeneration of the epidermis begins when a basal cell divides, resulting in occupation of two basal positions by the daughter cells. The next step is a migration into the prickle cell layer. The cells flatten, become granulated, and lose their organelles and nuclei as they move outward through successive strata to become keratinized squamous cells (corneocytes), finally flaking off from the skin surface as squames.

The different layers of the epidermis all contain keratin, but during the process of keratinization, different types of keratin are produced. These are composed of 19 different α-helix proteins with molecular weights ranging from 60,000 to 68,000 [21].

In human skin, the normal epidermal cell cycle is 20 to 30 days in duration, depending on the region of the body. The regeneration of the epidermis is regulated according to its thickness. Faulty control of the rate of proliferation causes a skin disorder called psoriasis in which the rate of basal cell proliferation is greatly increased, and epidermal cell cycles are completed within a week, without a complete keratinization.

Interspersed with the cells destined for keratinization, the so-called keratino-cytes, the epidermis contains small numbers of macrophage-like Langerhans cells, melanin-producing melanocytes, and neural Merkel cells [26].

III. DERMIS

Beneath the epidermis is the dermis. Unlike the epidermis, it contains blood vessels, an intercellular matrix, and fluid and lymphatic vessels that provide nutrients and immunological protection stemming from the reticuloendothelial system with its associated B- and T-cells [23]. The human dermis consists of two layers, the bound-aries of which are rather indistinct—the outermost papillary layer and the deeper reticular layer. The papillary layer is thin and composed of loose connective tissue, and the reticuloendothelial system is composed of fibroblasts, B- and T-lymphocytes, mast cells, macrophages, and extravasated leukocytes. Collagen fibers from the papil-lary region extend through the basal lamina and into the epidermis. They are thought to have the special function of binding the dermis to the epidermis as anchoring fibers. Intercellular substances provide strength and support of tissue and act as a medium for the diffusion of nutrients and metabolites between blood capillaries and the dermal cells that support the cellular metabolism. The papillary layer adheres tightly to the basement membrane of the epidermis, while the lower surface merges gradually with the reticular layer.

The reticular layer is thicker than the papillary layer and contains varying amounts of fat, often as a function of an individual's sex and age and the ana-tomical region of the body. Females generally show a greater amount of fat within the subcutaneous layers than do males. The reticular layer also contains the larger blood vessels and nerves, from which arise various superficial cutaneous branches to supply the skin [26, 27].

Both dermal layers are interspersed with connective tissue consisting of bundles of collagenous, elastic, and reticular fibers grounded in an amorphous, intercellular substance containing various types of histologically discrete cells. The dermal fibers add strength and elasticity to the skin, as well as accounting for its patterning and cleavage lines. In the papillary region, the bundles of collagenous and elastic fibers tend to be widely separated. In the reticular layer beneath, however, lies a dense network of coarse collagenous fiber bundles situated more or less parallel to the skin surface.

At least three important, histologically distinct cell types are found in the dermis [22, 23, 26]. The first, the fibroblast, is an undifferentiated cell capable of trans-forming into other connective tissue cell types. Another is the histiocyte of the reticuloendothelial system, a cell capable of phagocytosing inorganic or organic particles, including bacteria, viruses, and fungi. Histiocytes have been observed to unite in certain situations to form macrophagous giant cells able to phagocytose relatively large particles. The third important cell commonly found in the dermis is the Langerhans cell, which is similar to the monocyte in immune function. Morphologically, Langerhans cells resemble the fibroblast and are observed in the upper papillary layer, especially around capillaries, where they are often arranged in concentric rings.

The dermis is normally protected by the overlying, tough, keratinized epidermal covering, which prevents the entry of pathogenic microorganisms. However, when the epidermis is compromised, and microorganisms penetrate to the dermis or beyond, they may elicit both inflammatory and immune responses from the mononuclear phagocyte system (reticuloendothelial system) in normal, healthy individuals.

IV. DERMAL VASCULARIZATION

An important component of the dermis is its vascular system, which is composed of a rich network of blood and lymph vessels. Arterial blood brings oxygen and nutrients to all dermal cells, which utilize them at capillary junctions. The venous blood carries away products of metabolism collected at the capillary junctions. Additionally, the vascular system carries cells of the reticuloendothelial (immune) system to all parts of the dermis [21, 22, 23].

Figure 2.3 presents a diagrammatic representation of the various cell structures found in normal skin tissue. The skin is supplied by small arteries (arterioles) embedded in the subcutaneous layers. The arterioles form branches that pass outward to form the hypodermal arterial plexus. Branches from this plexus provide the vascular supply of hair follicles and sweat glands. Others pass through the dermis and branch to form dermal and subepidermal plexuses that give rise to many short vessels. The subepidermal plexes, which lie immediately below the basement membrane of the epidermis, form the arterial arms of the capillary loops that project into

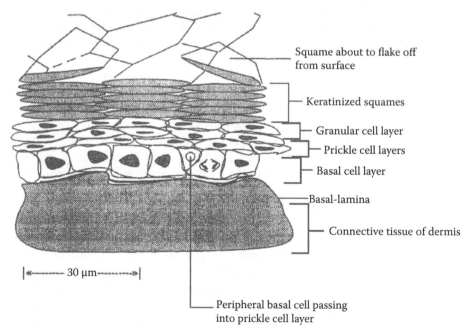

Squame about to flake off from surface

Keratinized squames

Granular cell layer

Prickle cell layers

Basal cell layer

Basal-lamina

Connective tissue of dermis

|◄———— 30 μm————►|

Peripheral basal cell passing into prickle cell layer

FIGURE 2.3   Cell structures of normal skin tissue.

the dermal regions. The blood passes through the arterial capillary loops into the wider venous arms and collects in a series of venous plexuses embedded at various levels in the dermis. Finally, blood reaches the hypodermal venules and passes to the subcutaneous veins. The arterial network is especially rich in the palms, soles, and the face, the vessels being both larger and more numerous than elsewhere. The cutaneous/cartilaginous appendages (e.g., ears, nose) also have rich vascular supplies.

Because the capillary loops have very small diameters approximating that of a red blood cell, temporary blockage of a capillary may occur. Further, the flow of blood through the capillary field is intermittent and may stop and start. There is also a constant variation in the number of capillaries open at any one time, since the opening and closing of capillaries is one of the mechanisms by which heat loss from the body is controlled. Despite the numerous plexuses supplying the skin, the vascular surface available for the exchange of oxygen and nutrients between the blood and the tissues is smaller than that of organs such as muscle. The dermis contains various structures of epidermal origin, namely, the hair follicles and the sweat and sebaceous glands, which are more or less numerous, depending on the region of the body. Also, a rich supply of nerves is found throughout the dermis [21, 22, 23].

A.   SUBCUTANEOUS TISSUE

This layer consists of diffuse connective tissue that binds the skin loosely to the subjacent organs. The overlying hypodermis often contains fat cells, varying in number and in size as a function of the area of the body, gender, and the nutritional state of the individual.

B.   HAIRS

Hair shafts are thin, keratinized structures derived from invagination of the epidermal epithelium. Their color, size, and disposition are variable according to race, age, sex, and the region of the body. Hairs grow discontinuously and exhibit periods of growth followed by periods of rest.

C.   SEBACEOUS GLANDS

Sebaceous or oily secretion glands are found embedded in the dermis over all areas of the body, except in those areas lacking hairs. These glands open into short ducts that end in the upper portion of a hair follicle. Sebaceous glands arise structurally from undifferentiated, flattened epithelial cells. The cell nuclei gradually shrink, and the cells become filled with fat droplets and burst. As these cells differentiate, sebaceous glands move toward the surface of the skin. The sebaceous gland secretes sebum, together with remnants of dead cells. Sebum is composed of a complex mixture of lipids containing triglycerides, free fatty acids, cholesterol, and cholesteryl esters and is secreted continuously. A disturbance in the normal secretion and flow of sebum is one of the reasons for development of acne [27].

D. SWEAT GLANDS

The sweat glands are designed for warming/cooling the body and are widely distributed in the skin in the form of simple, coiled, tubular structures. The fluid secreted by these glands contains mainly water, sodium chloride, urea, ammonia, uric acid, and proteins. The constant evaporation of sweat requires heat, which is withdrawn from the capillaries surrounding the gland; this loss of heat contributes to the thermoregulation of the body [26, 27].

V.   THE TISSUE FLUIDS

Because there is no vascular network residing in the epidermis, nutrients and oxygen for epidermal cells must diffuse varying distances from the blood capillaries. Although the capillaries do not appear to be actively contractile, the endothelial cells of which they are composed have an elastic property that makes them resistant to distortion, so that changes in tone result in changes in capillary permeability. Under normal conditions of intracapillary pressure, there is a tendency for fluid to flow from the capillaries into the tissue spaces on the arterial side of the capillary loop and to be reabsorbed from the tissues on the venous side, thereby providing a constant circulation. The action of the muscular walls of the feeder arterioles and of the precapillary sphincters can produce great variations in capillary pressures [22].

3 Skin Microbiology

The second leg of the skin–antimicrobial triangle deals with microorganisms.

Bacteria, fungi, and viruses are the causative agents of disease. In evaluating topical antimicrobial products such as surgical scrubs, preoperative prepping solutions, and healthcare personnel handwashes, bacteria, fungi, and viruses are also the indices for estimating antimicrobial effectiveness (see Figure 3.1).

I. ETIOLOGY OF INFECTIOUS DISEASES

For an infectious disease to spread, the following events must occur [28]:

1. *Encounter:* The host must be exposed to the microorganism (bacteria, fungi, viruses).
2. *Entry:* The microorganism must enter the host. For example, a fungus can attach itself to skin, nails, or hair. This is a superficial or cutaneous infection, where the fungus has been taken from its normal reservoir or soil. Other types of fungi can change from multicellular conidia-bearing mycelium into single-cell yeasts (dimorphisms) that can live in a parasitic form at 37°C.
3. *Spread:* The microorganism must spread from the entry site.
4. *Multiplication:* The microorganism must multiply within the host.
5. *Damage:* The microorganism and its metabolites and/or the host's immunological response cause light to extensive tissue damage in the host.

All five of these steps are required in breaching the host's defenses. The actual disease produced by the microorganism depends upon the entry conditions, the host's response, and the species and strain of microorganism [29].

A. ENCOUNTER AND ENTRY

A host is exposed to the microorganism (bacteria, fungi, viruses) in a variety of situations relevant to the user and type of topical antimicrobial compound [14]. For example, there is (1) the exposure of a host to pathogenic microorganisms via a healthcare/food-handling worker or by a different patient, worker, or environmental source; (2) the exposure of a patient's surgical site, a worker's wound site, or infection (e.g., nasal discharge with methicillin-resistant *Staphylococcus aureus* [MRSA]) that contaminates the hands; and (3) the contamination of a surgical, catheter, or injection site with the patient's own normal microbial flora, such as *Staphylococcus epidermidis*. Although reference may occasionally be made to food handler-transmitted infection, food-borne disease phenomena will not be discussed in detail. Rather, our focus will be nosocomial (hospital-acquired) and iatrogenic (physician-associated) infection.

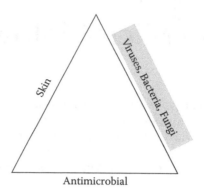

FIGURE 3.1   Skin–product–microorganism relationship.

In Case 1, the patient is exposed to infectious microorganisms from a different patient via the hands of contaminated healthcare personnel [30]. That is, the healthcare personnel serve as a disease vector, transferring patient A's pathogenic microorganisms to patient B. It should be noted that the microorganisms transferred are transient, not normal microorganisms of the skin.

Case 2 represents the exposure of a patient's surgical site to surgical personnel's contaminated hands. This often can occur when a surgical glove is torn or nicked, and the normal resident microorganisms (e.g., *Staphylococcus epidermidis*) of the surgeon's skin enter the surgical site of the patient [30]. The degree of resulting disease is dependent upon the surgical site, the number of contaminative microorganisms transferred to the patient, and the immunological competence of the patient.

Case 3 occurs as a result of contamination of a surgical site, venous/arterial catheter site, or an injection site with the patient's normal skin microorganisms [30, 31]. Contaminative surgical and catheter site infections are not uncommon and are dangerous because entry of microorganisms is often directly into the bloodstream, resulting in a bacteremia and potential septicemia. Patients who are immunocompromised are at a greater risk of septicemia. Injection site infections generally are localized, but can progress to a septicemic condition.

B.   SPREAD

Pathogenic microorganisms spread from patient to patient via hand contact can take a variety of forms. It can be hand-to-hand contact, such as shaking hands. Often, the contaminating microorganisms are respiratory or gastrointestinal in origin, but not exclusively. Most anatomical sites, particularly if they are physiologically compromised, can become a site of spread [28, 32].

Systemic infection resulting from direct hand contact with surgical personnel or introduction of normal skin microorganisms through catheters is due to the blood and lymphatic systems' dissemination. Such spreading can be passive or active, depending upon a microorganism's motility and/or manufacture of extracellular lytic enzymes that allow them to breach "walling off" mechanisms of the inflammatory response [28]. For example, surgical wound infections may be caused by

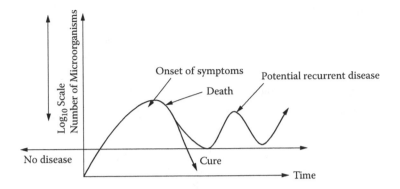

FIGURE 3.2   Incubation time.

Staphylococcus aureus. It produces a protease that breaks up fibrin, a hyaluronidase that hydrolyzes hyaluronic acid (an important component of connective tissue), and a deoxyribonuclease that causes the release of DNA from lysed white cells and reduces the viscosity of pus [33]. In addition, many strains of *Staphylococcus aureus* are resistant to the penicillins and other classes of antibiotic, presenting a major problem in disease treatment. Because the *Staphylococci* and other bacteria have demonstrated great ability to develop resistance to new drugs, the healthcare industry is challenged to stay ahead of these resistances.

C.   MULTIPLICATION

Once microorganisms have spread from the entry site, they frequently multiply to cause a systemic infection [14, 34]. Generally, there is a time lapse between the exposure and the symptom manifestation, referred to as the *incubation time* (Figure 3.2).

D.   IMMUNE SYSTEM

The pathogenic microorganisms' ability to multiply within the body is controlled by the body's immunological system, which is responsible for destroying them. In normally functioning immune systems, including the T- and B-lymphocytes, neutrophils, and macrophages, the body is able to generate a large variety of immunologically active cells that are capable of detecting (recognizing) and eliminating an apparently infinite variety of foreign substrates and microbial forms, including viruses, bacteria, and fungi [35, 36].

Functionally, the immune system can be divided into two interrelated activities: recognition and response. Immunological recognition is so specific that it is able to distinguish one pathogenic microorganism from another and to discriminate between *self* and *not self* at the molecular cell structure level. Once a foreign microorganism or foreign protein is recognized, the immune system mounts an immune response through the various immunological mechanisms—T- and B-cells, phagocytic cells, and certain molecular substrates [37]. Figure 3.3 depicts a schema that seems universal. First, the immune system differentiates a microorganism through

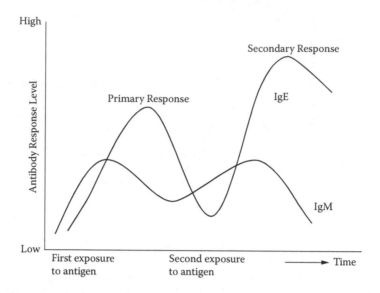

FIGURE 3.3 Primary and secondary immune responses.

its system's registry of antibodies. The first to appear in response to an inducer is Immunoglobulin M (IgM), followed by a relatively slow increase in levels of Immunoglobulin G (IgG). Frequently, these antibodies will clear the invader. When the inducer is encountered a second time, the IgM responds as it did previously, and the IgG responds rapidly and at increased levels. The response may then eliminate or neutralize the infectious substance. The immune system is able to translate the initial exposure to subsequent exposures because certain immunoactive cells retain their "memory." This phenomenon is the basis for vaccinations, which eliminate an invader rapidly and prevent disease.

Unfortunately, an individual who has been sick for some time or has undergone a surgical procedure is often immunologically compromised [38, 39]. The possibilities of morbidity and mortality through infectious disease are therefore heightened.

1. Microbial Nutrition

It would appear that the body offers a variety of rich mediums for microorganism support. Body fluids, including plasma and serum, contain sugars, vitamins, minerals, and other substances on which the microorganisms can subsist. For microorganisms other than normal flora, life in or on the body is not so simple. For example, if fresh blood plasma is incubated with challenge microorganisms, microbial growth is generally nonexistent or sparse. This is because antimicrobial substances such as lysozyme inhibit constituents of the immune system [40, 41].

Additionally, bacteria require free iron for the synthesis of their cytochromes and other enzymes. Plasma, as well as a variety of other body fluids, contains very little free iron, probably due to its binding to a wide range of proteins [42]. In fact, the body actually sequesters iron in defense against bacterial multiplication [28]. When a sufficient number of microorganisms have been detected by the immune system,

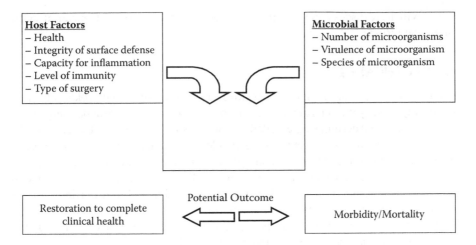

FIGURE 3.4   Physical factors.

iron-binding proteins literally pour into plasma and other tissue fluids to limit the availability of free iron to bacteria.

The nutritional range requirements of microorganisms that constitute the normal microbial flora are a reflection of their ecological niches. For example, *Staphylococcus epidermidis,* the predominant skin surface-colonizing bacterium, requires several amino acids and vitamins that are found commonly on the skin surfaces [28]. However, microorganisms common in both soil and water are much less fastidious. They can achieve their organic requirements from simple carbon compounds widely available in the body, as well as the natural environment. Both *Escherichia coli* and *Pseudomonas* spp. are examples of bacteria that can thrive on very nutrient-minimal media [43].

2. Physical Factors

Physical factors also affect microbial multiplication within the body. These include temperature range of the body's anatomical region, osmotic pressure of fluids, and humidity [44]. Microorganisms that are normal inhabitants of the body, on skin or within the gastrointestinal tracts, and various other orifices, tend to have limited tolerance to physical changes that are found commonly in the environment. Figure 3.4 presents a diagram of the physical factors that are important.

3.   Endogenous Microorganisms of the Body

It is important, in studying the role of topical antimicrobial products, to be familiar with the microorganisms that comprise the skin's normal flora. These are important, particularly in the surgical environment where such species may be introduced into a surgical wound by surgical staff, or contaminatively from the patient's own skin, to produce an opportunistic infection.

Normally, the body is colonized by many different bacterial species, as well as viruses and fungi [29]. The most common sites of colonization are aspects of the digestive tract (mouth and large intestine), the respiratory tract (nose and oropharynx), the female genitalia, and the skin surfaces [41].

We will next discuss specifically the structure and function of bacterial cells, fungi, and viruses.

II. BACTERIAL CELL

Two basic cell types are found in most life forms: eurkaryotes and prokaryotes [32, 44, 45, 46]. Eukaryotic cells are unit structures of higher plants, animals, fungi, most kinds of algae, and certain single-celled organisms such as protozoa. Eukaryote cells are characterized by true nuclear bodies (nucleus) bounded by a nuclear membrane and containing chromosomes that undergo mitotic division (and meiotic division in germinal cells) [32]. They also possess various cellular organelles, such as the Golgi body and endoplasmic reticulum, that function in sorting and transporting proteins destined for secretion into the plasma membrane or lysosomes; the mitochondrion, both of which play critical roles in energy metabolism; and the ribosome, a structure responsible for generating peptides and proteins [47].

Prokaryotes are bacteria and blue-green algae that are structurally simpler than the eukaryotes. They do not contain a nucleus bounded within a membrane, but instead have a simple nuclear filament, replicated nonmitotically [47, 48].

Prokaryotic cells, with a few exceptions, are enclosed within a rigid cell wall containing a unique constituent, muramic acid. Bacteria are classified morphologically into one of three categories: cocci (spheres), bacilli (rods), or spirals [49]. The cocci and bacilli are the most commonly observed bacteria among human pathogens [48]. Bacteria are distinguished further as being Gram-positive or Gram-negative, based on the Gram staining method [48]. Devised by Hans Christian Gram, the Gram stain ranks among the most important stains for identifying bacteria, allowing one to distinguish among bacteria that exhibit similar morphology [41]. Gram-positive bacteria, following an acetone-alcohol destaining step in the procedure, retain a crystal violet-iodine complex and appear purple [48]. Other bacteria do not retain the complex when destained, but are then stained red by the safranin dye counterstain step that follows.

Using this procedure, one can determine microscopically the bacterial size, form, and Gram-positive or Gram-negative property.* There are a number of other staining techniques useful in the study of bacteria, including acid-fast staining, capsule-staining, flagellar staining, metachromatic granule staining, spore staining, and relief staining [48].

While the great majority of bacteria falls into one of the three basic morphologies (coccus, bacillus, spiral), they may exhibit variations on, or blending of, shapes in older cultures, in which cell structures become weakened and break down [50]. For example, cells may appear balloon-like or granular. Such degenerative forms result from the breakdown of mechanisms that control selective permeability and

---

* The difference between Gram-negative and Gram-positive can be attributed to differing chemical constituents in the cell wall. Gram-negative bacterial cell walls have a higher lipid content than do Gram-positive bacteria. Although a crystal violent-iodine complex is formed in both kinds of cell, the acetone–alcohol leaches the lipid from the Gram-negative bacterial cells, which increases the permeability, resulting in loss of the dye complex. The complex is retained, however, in the Gram-positive cell walls, in which dehydration by the acetone–alcohol causes a decrease in permeability.

from enzymatic autolysis. Aberrant forms may also be produced by the culturing bacteria under adverse environmental conditions, including higher than optimal temperature ranges, the presence of high concentrations of inorganic salts, or exposure to sublethal levels of antimicrobial products. Finally, some species are pleomorphic, exhibiting multiple morphologies; some are Gram-variable; and some lack a cell wall and do not Gram-stain at all.

A. BACTERIAL CELL STRUCTURES

Most bacteria are enclosed in multilayered structures that include, from inside to out, a cytoplasmic (plasma) membrane, a cell wall with associated proteins and polysaccharides, and in some, protective capsules and slime layers that serve to protect against the host's immune system, particularly phagocytosis [28, 44, 47]. Many also bear external filamentous appendages (flagella and pili) that function in locomotion, tissue attachment, or specialized reproductive processes.

In general, the bacterial cell wall is a rigid structure that encloses and protects the inner cellular contents, the protoplast, from physical damage (e.g., conditions of low external osmotic pressure). The rigid cell wall is thought to be an evolved structure that allows bacteria to tolerate a wide range of environmental conditions. The protoplast, a bacterium less the cell wall, is composed of a cytoplasmic membrane and its internal contents.

Internally, bacteria are relatively simple prokaryotic cells [48, 49]. Major cytoplasmic structures include a central fibrillar chromatin network surrounded by an amorphous cytoplasm containing ribosomes. Cytoplasmic inclusion bodies—energy storage granules—vary in chemical composition, depending upon the bacterial species, and in number, dependent upon the bacterial growth phase and environmental conditions (e.g., temperature). Some cytoplasmic structures (e.g., endospores) are produced by only a few bacterial species. Typical Gram-positive and Gram-negative bacterial cells, which differ primarily in cell wall organization, are shown in Figure 3.5.

1. Bacterial Size and Form

Medically important bacterial species vary from approximately 0.4 to 2 μm in size and appear under the light microscope as spheres (cocci), rods (bacilli), or spirals (vibros or spirochetes) [32, 44]. Cocci are found singly, in pairs as diplococci, in chains (e.g., *Streptococcus* spp.), or, depending upon division planes, in tetrads or in grape-like clusters (e.g., *Staphylococcus* spp.). Bacilli vary considerably in length and width, from very short rods (coccobacilli) to long rods that can measure several times their diameter. The terminal ends of bacilli may be gently rounded, as in enteric organisms such as *Salmonella typhi,* or squared, as in *Bacillus anthracis.* Long sequences of bacilli are called chains, while long, thread-like bacilli are generally referred to as filaments. Fusiform bacilli, found in the oral and gut cavities, are tapered at both ends [47]. Curved bacterial rods vary from small, comma-shaped or mildly helical forms with only a single curve, such as *Vibrio cholerae,* to longer spiral (spirochete), multiply-coiled forms, such as species of *Borrelia, Treponema,* and *Leptospira.*

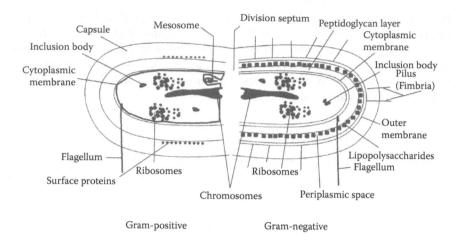

FIGURE 3.5   Cross sections of Gram-positive and Gram-negative cells.

B.   CELL ENVELOPE

Bacterial cells are bounded by the cell envelope, an integrated structure of varying complexity and generally of several layers [32, 44, 50]. The bacterial cell envelope commonly includes a cytoplasmic membrane, an overlying cell wall, certain proteins and polysaccharides, as well as various outer adherent materials. This multilayered structure constitutes 20% or more of the prokaryotic cell's dry weight [45]. The bacterial cell envelope contains transport sites for nutrients, receptor sites for specific bacterial viruses, sites of host antibody and complement reactions, and components potentially toxic to the host.

The occurrence of bacteria in shapes other than spheres (cocci) is evidence that envelope rigidity is sufficient to withstand environmental surface tension and internal turgor of the cell. This rigidity and the resulting cell shape are largely attributable to the cell wall component, which also provides resistance to mechanical disruption.

Among bacteria, the cell envelope varies considerably in structural complexity and differentiation. The simplest envelopes are found among the members of the genus *Mycoplasma*, which possess only a cell membrane, which averages about 7.5 nm thick [47]. Because the cell membrane structure has the dual role of *cell wall* and *cytoplasmic membrane* and does not contain the peptidoglycans found in true cell walls, mycoplasmas possess limited cellular strength.

The cell envelope of Gram-negative bacteria is the most complex (Figure 3.5). The inner layer of the envelope is a cytoplasmic membrane, structurally similar to that of Gram-positive cells. The overlying cell wall is composed of a second (outer) membrane often appearing wrinkled in appearance with a thick peptidoglycan layer [49]. The Gram-negative wall is thinner than that of Gram-positive species, usually averaging about 10 to 15 nm across.

1.   Capsules and Slime Layers

The virulence—the ability to cause disease—of pathogenic bacteria is often related directly to their production of a capsule [28, 29, 32]. For example, virulent strains of *Streptococcus pneumoniae* produce capsules that protect them from phago-cytosis by neutrophils and macrophages. It has been demonstrated that loss of the capsule-forming ability produces a loss of virulence and increased ease of destruction by host phagocytes. While capsules and slime layers are composed of similar *gels* that adhere to the outer cell wall, slime layers consisting of extracellular polymers are more easily *washed off* than capsules [47].

Most bacteria, possibly all species found in the environment (wild types), are surrounded by a layer of the gelatinous, poorly defined and poorly stainable mate-rial that has been termed the *capsule, slime layer*, or *glycocalyx*. The term *capsule* applies to the material surrounding a single cell, while the *slime layer* is used to describe the matrix that envelopes a microcolony or group of bacterial cells, consis-tent with the term *biofilm*, to be discussed later [47].

The capsule enclosing the cell walls of some bacteria had been thought by some to be simply the waste products of cellular metabolism and not of great importance to bacteria [48]. This view, for the most part, has been effectively challenged [47]. It is clear that capsules play a key role in microbial ecology, not solely in the host–parasite relationship, but also in a species' growth and survival in the natural environment. Bacteria, whether derived from water or soil or from the mucus, blood, or tissue of a host, nearly always possess a capsule. Only after an in vitro cultivation in the labora-tory setting are the capsules of many species lost or diminished.

The capsules surrounding individual bacterial cells or the slime layers surround-ing the bacterial microcolonies serve several known purposes [45, 46, 47]. They protect bacteria from desiccation, bacteriophage (virus) penetration, and toxic heavy metals. The anionic nature of capsules also attracts nutrients from the surrounding environment. Finally, capsules allow bacteria to adhere to the cells of hosts and vari-ous substrates in the natural environment, permitting colonies to grow.

The size of the capsule varies widely among bacterial species. In heavily encapsu-lated forms such as the pneumococci and *Klebsiella* spp., the thickness of the capsule is frequently greater than the diameter of the bacterial cell. In other bacteria, it may be but a thin layer, closely adherent to the cell's surface.

Bacterial capsules are not stained by usual procedures such as Gram- or acid-fast staining, because the capsule layer fails to retain these dyes [48]. Capsules may be demonstrated under the light microscope using negative staining techniques, such as suspending the bacteria in diluted India ink. The capsule displaces the colloidal carbon particles in the ink, causing the cells to appear to lie in lacunae against a dark background (Figure 3.6).

2. Constitution of Capsular Substances

The capsule is mainly composed of polysaccharides, varying in both complexity and specific composition, and appearing in both linear and branched chains. Chemically, the simplest consist of homopolysaccharides, which are polymers of a single type of monosaccharide [44, 45]. Examples include bacterial celluloses, levans, dextrans, and glucans. Most capsules, however, consist of the more complex heteropolysaccha-rides, often with uronic acid as an additional component [41]. The monomeric con-stituents of these polysaccharides may include D-glucose, D-mannose, D-glactose, L-fructose, L-rhammose, ribitol, and glycerol, as well as various amino acids and

FIGURE 3.6   Suspended bacterial cell.

uronic acid [45, 47]. Phosphorus may also be present, particularly in polysaccharides containing polyols (e.g., ribitol and glycerol) and teichoic acids [49].

The biosynthesis of most polysaccharides takes place in the cell at the cytoplasmic membrane, involving diphosphate sugars, as well as isoprenoid lipid intermediates. The polysaccharides are then transported to the extracellular location as capsule material. In some species, however, capsule synthesis is less complex. The sugars (e.g., dextrans, levans, etc.) are formed by an extracellular process that does not involve the nucleoside diphosphate sugars or the lipid carriers. And, in other species, notably *Bacillis* sp., the capsule is not a polysaccharide but a polypeptide [44].

The cell wall is found in all bacteria except *Mycoplasma* spp. The cell wall, being rigid, protects the cell from bursting in low osmotic pressure environments and maintains cell shape [45].

a.   Gram-Positive Cell Envelope

In association with a peptidoglycan structural component, Gram-positive bacteria produce specific surface polysaccharides and proteins. The better-known polysaccharides include teichoic acids, comprising many of the pneumococcal capsular substances and streptococcal group polysaccharides [45]. Poly-D-glutamic acid polymers are produced by some *Bacillus* species, and the M-protein of Group A streptococci is a virulence factor. Observed under electron microscopy, thin cross sections of Gram-positive cells reveal a relatively thick, contiguous cell wall layer overlying the plasma membrane. Both proteins and polysaccharides contribute to the layered wall substructure. The Gram-positive cell wall is sensitive to and can be destroyed by lysozyme, an enzyme common in skin secretions and tears.

b.   Gram-Negative Cell Envelope

Not including a capsule, Gram-negative bacteria exhibit three layers in their cell envelopes (Figure 3.5). These are the outer membrane, a middle dense layer, and the inner cytoplasmic (plasma) membrane [32, 44]. The middle dense layer between the outer and cytoplasmic membranes—termed the *periplasmic space*—is occupied by a gelatin-like structure, bounded by a layer of peptidoglycan lying beneath the outer membrane. Both the plasma and outer membranes appear as a trilayered sandwich structure, when observed using transmission electron microscopy. A helical lipoprotein, one third of which is covalently linked at one end to the peptidoglycan layer, inserts its lipid end into the outer membrane, thereby anchoring the outer membrane to the cell [49]. The cell envelope can be isolated free of soluble cytoplasm by cell rupture and differential centrifugation. The inner membrane can be dissolved with mild nonionic detergents, leaving the outer membrane bound to insoluble peptido-glycan. The outer membrane can be disrupted by ethylenediamine tetraacetic acid (EDTA), strong ionic detergents, aqueous phenol, or butanol extraction.

c. Outer Membrane

Gram-positive species do not exhibit an outer membrane, per se. The outer membranes of Gram-negative species are similar in appearance and contain lipopolysaccharide (also known as the *somatic O surface antigen* or *endotoxin*), phospholipids, and unique proteins that differ from those found in the cytoplasmic membrane. Compared to the cytoplasmic membrane, the outer membrane contains less phospholipid and fewer protein compounds, but contains unique bacterial lipopolysaccharides [46]. The inner and outer leaflets of the outer membrane are also uniquely asymmetrical, appearing in electron micrographs as two electron-dense layers sandwiching an electron-neutral (translucent) middle layer. The outer leaflet is functionally distinct for most Gram-negative bacteria. It acts as a hydrophobic diffusion barrier against a variety of substances and contains receptors for both bacteriophages (viruses) and bacteriocins. The outer leaflet has a participatory role in both cell division and con-jugation and also contains a number of systems mediating the uptake of nutrients and the passive diffusion of small molecules into the periplasmic space. The outer membrane, in conjunction with the thin peptidoglycan layer, provides structural integrity. In Gram-negative enteric bacteria (e.g., *Salmonella),* phosphatidylethanol-amine occurs almost exclusively in the inner leaflet, whereas the anionic, hydrophilic lipopolysaccharide occurs only in the outer leaflet. In *Neisseria* and *Haemophilus* species, however, phospholipids are found in both the inner and outer leaflets of the outer membrane [46]. Proteins known as porins are distributed throughout the outer membrane and form transmembrane diffusion channels for low-molecular-weight, water-soluble substances, or serve as bacteriophage (virus) receptors.

d. Lipopolysaccharide

Lipopolysaccharides are part of the outer membrane and are unique to Gram-negative bacteria [51]. They comprise the major surface antigenic components of the somatic O antigens and are responsible for the endotoxin properties of Gram-negative bacteria cells, in general [44].

Lipopolysaccharides are high-molecular-weight complexes usually consisting of three regions: a lipid region, termed Lipid A; a core polysaccharide region; and a specific polysaccharide region, the O-specific chains (Figure 3.7) [44, 51].

The O-specific chains project from the surface of the outer membrane and serve to protect the cell from antibodies, and complement by interfering with their direct

| Lipid A | Core Region | O-Specific Side Chains |

FIGURE 3.7   Basic lipopolysaccharide structure of Gram-negative bacteria.

contact with the outer membrane [32]. The Lipid A region is responsible for the endotoxin activity of bacterial lipopolysaccharides and is situated in the hydrophobic zone of the outer membrane.

e. Outer Membrane Proteins

Many of the outer membrane properties previously described are attributable to the proteins found in the membrane. Those proteins present in the greatest amounts are designated as the *major* outer membrane proteins and are divided into three groups: (1) the pore-forming proteins (porins), (2) the nonporin proteins, and (3) the lipoproteins [52].

The proteins of the Gram-negative outer membrane act to permeate solutes through the membrane [26]. The simplest, the matrix porins, form transmembrane channels through which hydrophilic solutes of low molecular weight flow. Solutes do not interact with the porins in this passive permeation. Other porins have more specific permeation functions, such as interacting with the substrate as a part of its membrane translocation.

The proteins are not well understood as to their role as porins. Some are known to span the outer membrane and associate with the peptidoglycan layer. They may serve as receptors for a variety of ligands, regulate exopolysaccharide biosynthesis, or help stabilize the cell envelope [52] (Figure 3.8).

The lipoproteins are the smallest and the most common of the outer membrane proteins. Lipoproteins are usually covalently bound to peptidoglycan and are essential to structural stabilization of the cell envelope [32]. The role of those lipoproteins shown to be free and unbound is not clearly understood.

f. Gram-Positive Cell Wall Structure

Electron microscopy of thin sections of Gram-positive bacteria demonstrate an amorphous, electron-dense layer lacking fine structure throughout. In Gram-positive bacteria, the cell wall is 15 to 50 mm in thickness, making up 20 to 40% of the dry weight of the cell. The cell wall of Gram-positive bacteria is composed, to a large degree, of 20 to 50% of peptidoglycan layers [45, 52].

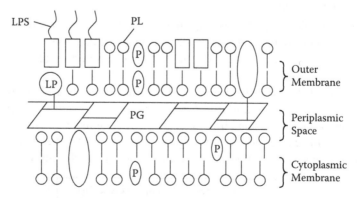

FIGURE 3.8 Gram-negative cell envelope (LPS: lipopolysaccharide; LP: lipoprotein; P: protein; PG: peptidoglycan; PL: phospholipid).

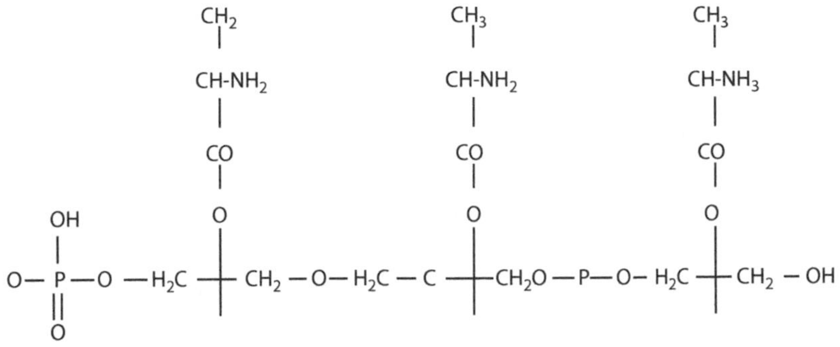

$$CH_2 \qquad\qquad CH_3 \qquad\qquad CH_3$$
$$CH\text{-}NH_2 \qquad CH\text{-}NH_2 \qquad CH\text{-}NH_3$$
$$CO \qquad\qquad CO \qquad\qquad CO$$

OH $\qquad$ O $\qquad\qquad$ O $\qquad\qquad$ O

O— P —O —H_2C —|— CH_2 —O— H_2C — C —|—CH_2O —P—O— H_2C —|— CH_2 — OH

||
O

FIGURE 3.9 General teichoic acid structure.

Within the wall, occurring on its surface layers, are a variety of polysaccha-rides, proteins, and teichoic acids. Many of these surface components are immuno-logically reactive, particularly the surface proteins (C-polysaccharides) found on *Streptococcus* spp. Of this group, the teichoic acids are of special concern. They are constituents of the cell wall and membrane having a basic structure of glycerol phosphate or ribitol phosphate linked via phosphodiester bonds (Figure 3.9).

D-alanine and carbohydrate components may be ether-linked to this linear backbone structure. These teichoic acids occur in Gram-positive bacteria as membrane-associated (glycerol type) or wall-associated (glycerol or ribitol) types [45, 47, 52].

The membrane-associated teichoic acids are covalently linked to glycolipids contained in the plasma (cytoplasmic) membrane. These compounds are commonly termed *lipoteichoic acids*. The glycolipid portion of the molecule is attached to the outer leaflet of the membrane with hydrophilic polyglycerol phosphate chains extending outward toward the cell wall [46].

The wall-associated teichoic acids are covalently linked to peptidoglycan con-tained in the cell wall structure [50, 52]. Because these teichoic acids can participate in immunological reactions, they likely are partially exposed on the cell surface. The teichoic acids probably serve to stabilize the cell wall and maintain its contact with the cytoplasmic membrane. Teichoic acid is also thought to contribute to magne-sium binding and help maintain proper ionic range condition for cationic-dependent enzymes in the cell envelope. Because they behave as cationic-binding polymers, they can act as ion exchangers in a manner similar to polysaccharides in the slime layer or capsule production. Their presence on the bacterial cell surface may also enhance attachment of pathogenic Gram-positive bacteria to the mucosal surfaces [51] (Figure 3.10).

Although not usually found in isolated states, teichoic acids have been shown to induce specific antibody formation when combined with other constituents in the complete cell [53]. They are, therefore, important as surface antigens of many Gram-positive bacteria.

The topography of the outer surface of bacterial cells is of interest because it provides insight into localization of specific cell wall components. This section on electron microscopy has not been particularly informative, but freeze-etching and negative staining have revealed a regularly occurring array of macromolecules on

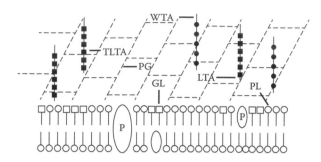

FIGURE 3.10 Gram-positive cell wall (LTA: lipoteichoic acid; TLTA: transient lipoteichoic acid; WTA: wall-associated teichoic acid; P: protein; PG: peptidoglycan; PL: phospholipid; GL: glycolipid).

the cell surface designated as the *S-layer* [48]. Most of these surface patterns in Gram-positive bacteria consist of subunit arrays having tetragonal symmetry, with center-to-center spacing ranging from 7 to 16 nm. These appear to be composed of single subunit types, usually protein, covering the entire surface of the cell.

g.   Adhesion Sites (Bayer Junctions)

In Gram-negative cells, points of connection between inner and outer membranes are known as adhesion sites, or Bayer junctions [52], and these are physiologically active [45]. On the outer membrane, they are sites of bacteriophage attachment–DNA injection and complement-mediated lysis. Internally, adhesion sites appear to be growth zones, serving as sites for translocation of secretory protein, outer membrane proteins, lipopolysaccharides, and capsular polysaccharides, and as emergence sites for pili and flagella.

h.   Periplasm

Periplasm, which occurs in the space between the cytoplasmic and the outer membranes, may be readily observed in Gram-negative bacteria, but not at all or only with difficulty in Gram-positive bacteria [44, 52]. This may be explained by the relatively high internal osmotic pressures of Gram-positive bacteria, compared to those of Gram-negative bacteria. The periplasmic gel space of Gram-negative bacteria varies with growth conditions and among individual bacteria. The gelatin, quite viscous and possibly highly ordered in structure, surrounds and is interspersed with the porous peptidoglycan. It contains cytoplasmic membrane-derived oligosaccharides that occur in inverse proportion to the osmolarity of the growth medium, various hydrolytic enzymes such as phosphatases, nucleases, plasmid-controlled β-lactamases (penicillinases), and proteins that specifically bind sugars and transport amino acids and inorganic ions. Release of these constituents from the periplasmic space can be induced by osmotic shock—that is, rapid dilution of hypertonic cell suspensions—after EDTA treatment [51].

i.   Cytoplasmic (Plasma) Membrane

Beneath the rigid cell wall layer, and in close association with it, is the delicate cytoplasmic membrane, which is vitally important to the cell. In thin sections, electron microscopy reveals the plasma membrane to be a typical trilaminar sandwich structure of dark-light-dark layers [32].

The bacterial cytoplasmic membrane is similar in chemical composition and structure to that of eukaryotic cells. Both are composed of a phospholipid bilayer with proteins interspersed in the membrane. The membrane is composed of 50 to 75% protein, 20 to 35% lipid, and in bacterial cells, comprises about 10% of the dry weight [45].

The bilayer membrane is divided by a hydrophobic zone, traversed by proteins that are thought to be permeases involved in the active transport of small substrates (e.g., various amino acids, carbohydrates, etc.) to the cell's interior. Cytoplasmic membrane proteins are also associated with both inner and outer layers of the membrane. Other proteins of the membrane are related to enzymatic functions, such as oxidative phosphorylation, macromolecular synthesis of the cell wall, and other tasks less understood [50, 52]. Within the cytoplasmic (inner leaflet) side of the membrane are situated enzymes responsible for various cytoplasmic maintenance functions.

The cytoplasmic membrane lacks the strength and rigidity of the bacterial cell wall. The membrane is incapable of maintaining the cell shape, other than that of a spherical coccus, if the cell wall is removed [50]. The membrane is held firmly beneath the cell wall by internal osmotic pressure. Hence, the cytoplasmic membrane is structurally distinct from the cell wall, but for all practical purposes, the two are chemically bonded.

j. Cytoplasmic Membrane as Osmotic Barrier

Although bacteria are regarded as extremely tolerant of osmotic changes in their external environment, their protoplasts (Gram-positive bacteria lacking cell walls) or spheroplasts (Gram-negative bacteria lacking cell walls) undergo either plasmolysis (shrinkage) or plasmoptysis (swelling) when placed in appropriate media. Cell wall–free bacteria are very susceptible to the vagaries of the environment. Placing intact bacteria into hypertonic solutions results in plasmolysis, that is, shrinkage of the membrane and cytoplasm from the cell wall. Gram-negative cells are more easily plasmolyzed than are Gram-positive cells, which correlates with their relative internal osmotic pressures.

The osmotic barrier is demonstrated by an ability to concentrate certain amino acids against concentration gradients which, in Gram-positive bacteria, may be as much as 300- to 400-fold [52]. Phosphate esters, amino acids, and other solutes contribute to the internal osmotic pressure. Osmotic activity is also indicated by selective permeability to various compounds.

3. Membrane Components

Membranes account for some 30% or more of the bacterial weight [32, 44] and comprise 60 to 70% protein, 30 to 40% lipids, and small amounts of carbohydrate. Phosphatidylethanolamines (75%), phosphatidylglycerol (20%), and glycolipids are found as major constituents. Choline, sphingolipids, polyunsaturated fatty acids, inositides, and steroids are generally absent [56]. Pathogenic *Mycoplasma* spp., however, do incorporate steroids from the environment into their plasma membranes.

Glycolipids, including glycosyl diglycerides, are found primarily in Gram-positive bacterial membranes, which also contain lipoteichoic acids.

Various enzymatic activities associated with membrane proteins include the energy-producing bacterial cytochrome and oxidative phosphorylation systems, membrane permeability systems, and polymer-synthesizing systems. An ATPase has been isolated from knob-like membrane structures similar to those found in eukaryotic mitochondria [52, 53]. Up to 90% of the cellular ribosomes exist as a membrane-polyribosome-DNA aggregate.

a. Mesosomes

The cytoplasmic membrane may be invaginated to form internal organelles, or the mesosomes [52]. These can assume a vesicular, lamellar, or tubular form, and more than one of these forms may be observed in a bacterial cell. They occur in all three forms in Gram-positive bacteria, but the laminar type is typically observed in Gram-negative bacteria. The gross chemical composition of the mesosome is equivalent to that of the cytoplasmic membrane, and although many functions have been proposed, none have been firmly established. However, mesosomes have been linked largely on morphological grounds with cell replication and apportionment of DNA to daughter cells in cell division, as they are known to act as cell division septum initiators.

b. Peptidoglycan Layer

The shape and rigidity of bacterial cells are almost entirely due to the presence of a polymeric support structure called peptidoglycan (also murein, mucopeptide, or glycosaminopeptide) that lies just outside of the cytoplasmic membrane [45, 52]. There are only a few species of bacteria that lack this structure, but in particular, the *Mycoplasma* spp., which lack cell walls in situ.

The peptidoglycan layer can be thought of as a single sack-like, macromolecular structure completely encompassing the cytoplasmic body of the cell. The peptidoglycan layer is a crystal lattice structure composed of glycan strands cross-linked with short peptide chains, and the strands themselves are composed of alternating units of (β-1, 4-linked N-acetyl-D-glucosamine and N-acetyl-D-muramic acid), the 3-lactylether of D-glucosamine [45, 52, 53]. The chain lengths of individual glycan strands vary from approximately 10 to 170 disaccharide units (Figure 3.11).

The N-terminus of the peptide subunit is bound through the carboxyl group of muramic acid, and the amino acid sequence of the linking peptide is usually L-alanine to D-isoglutamine to meso-diaminopimelic acid (or L-lysine) to D-alanine. These peptide subunits of adjacent glycans are directly joined through the unbound amino group of diaminopimelic acid and the C-terminus of D-alanine or indirectly through an interpeptide bridge. Although there is a remarkable degree of constancy in the glycan strands, different groups of bacteria show variation in the composition of the interpeptide bridges and, to a lesser degree, in the composition of the peptide chains [45].

The peptidoglycans are susceptible to the hydrolytic activities of a number of enzymes, many of which are produced by potential hosts for pathogenic bacteria. The most studied of these is lysozyme (N-acetylmuramidase), which hydrolyzes

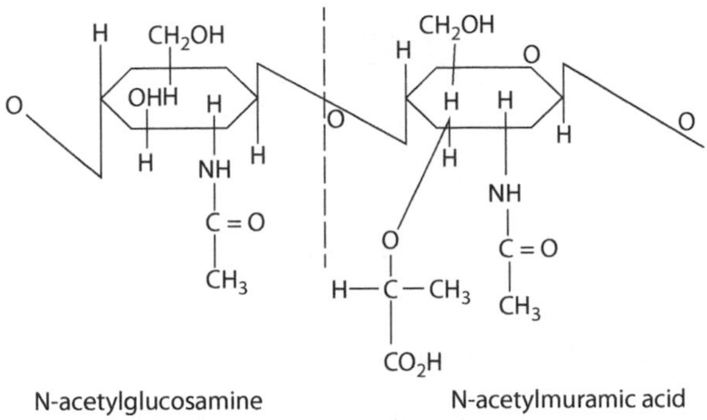

N-acetylglucosamine   N-acetylmuramic acid

FIGURE 3.11 Disaccharide structure of peptidoglycan.

the glycan strand at the glycosidic linkages between N-acetylmuramic acid and N-acetylglucosamine (Figure 3.11). Lysozyme and other such enzymes thereby depolymerize the peptidoglycan and, in Gram-positive bacteria in particular, destroy the cell wall, resulting in cell lysis [45, 52].

4. Bacterial Endospores

The endospore is an oval to spherical body formed within bacilliform bacteria that represents a dormant state highly resistant to lethal effects of heat, drying, cold, lack of nutrients, and many chemicals lethal to the vegetative forms of the bacteria [47, 49]. Its high resistance to lethality provides the spore formers with a survival advantage over nonspore-forming species of bacteria [54]. Endospore formation, however, is uncommon in bacteria and is limited to the species of *Bacillus* and *Clostridium*, both widely distributed in nature, plus a few minor genera of aerobic and anaerobic bacteria. Endospores are light-refractile, and their size, shape, and position in the mother cell are relatively constant characteristics of a given species [46]. Endospore coats include a rigid peptidoglycan layer, which differs in composition from that of the parent vegetative cell in that peptide cross-linkages are markedly reduced and large amounts of dipicolinic acid and calcium ions are incorporated [54]. Surface antigens of endospores usually differ from those of the vegetative cell [32, 44]. Additionally, because endospore formation represents a form of cell differentiation, it is of interest to cell biologists, as well as to specialists in sterilization technology who use the resistant spores as indicators of sterility [54].

Within the vegetative cell, the endospore is spheroidal or oval with the long axis parallel to that of the bacillus. Its breadth may be the same as the vegetative cell, or even greater, resulting in a bulging of the cell commonly seen in some *Clostridium* spp. The relative size of the endospore is constant, as is its position in the vegetative cell, which can be in the center of the vegetative cell (central), partway between the center and the end of the cell (subterminal), or at the end of the cell (terminal) [52, 54]. Hence, endospore size and location define, in part, the characteristic morphology of a species. For example, *Clostridium tetani* have a "drumstick" appearance because of the large terminal spore.

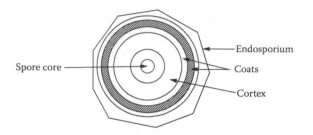

FIGURE 3.12   Diagram of a typical endospore.

Endospores can be observed in living bacteria using phase-contrast light micros-copy. In fixed (killed) preparations, endospores do not Gram-stain and require heat-ing for stain (usually carbolfuchsin) to penetrate them [48]. Endospore structure is complex [54]. The protoplast (spore core), the innermost portion, is bound by a limiting membrane analogous to the cytoplasmic membrane of the vegetative cell. A cortical layer (cortex) encloses the protoplast. The cortex, in turn, is enclosed by one or more complex cystine-rich protein layers termed the *spore coat(s)*, which are unique to spores, containing compounds not found in vegetative cells. Finally, the endospore may be surrounded by a thin, often loose-fitting bag-like structure, the exosporium. Figure 3.12 provides a diagrammatical representation of an endospore.

a.   Dormancy

Bacterial endospores are relatively inactive metabolically [54]. In many cases, there is no significant endogenous metabolism, but in some, minimal metabolic reactions are detectable. Even so, it is clear that the endospore has the metabolic capacity for both germination and vegetative growth when conditions are advantageous. Dormancy may be associated with the presence of large amounts of the unique chemical com-pound, dipicolinic acid, in the protoplast. Chelated with calcium ions, dipicolinic acid is also important in the structure of the endospore cortex, and intercalated within the helical structure of DNA, it displaces intramolecular water. The heat resistance of bacterial spores is thought to be due to the reduced amounts of water in the densely configured, greatly compressed core structure. Notably, dipicolinic acid is virtually absent from the vegetative cells prior to sporogenesis.

b.   Bacterial Nucleus

Using the Feulgen stain, bacterial DNA can be detected by light microscopy as nucleoids, or chromatin bodies [32, 44]. It is difficult to demonstrate chromatin bodies by direct staining because of the high concentration of RNA in the bacterial cell, but this can be removed by using ribonucleases. Chromatin bodies are present at all stages of bacterial growth.

Electron microscopy of thin sections of bacteria reveal that the nuclear material is an irregular, thick, fibrillar DNA network. Often running parallel to the axis of the cell, a direct attachment to the cell membrane can be clearly observed. During multiplication, bacterial DNA remains as a diffuse chromatin network and never aggregates to form a well-defined chromosome as is observed in the mitosis of eukaryotic cells.

Ribosomes

Examination of negatively stained thin sections of bacterial cells using electron microscopy reveals the presence of 70 Svedberg-unit (70S) ribosomes, each of which is comprised of 30 S and 50 S subunits [52, 53]. The ribosomes are composed of ribosomal RNA (rRNA). The 30 S subunit is responsible primarily for decoding the messenger RNA (mRNA) for protein synthesis. The 50 S subunit contains the 23 S and 5 S rRNA molecules, which are involved primarily with peptide bond formation in the assembly of amino acids into peptide chains [45]. The ribosomes tend to exist in aggregates of varying sizes called polysomes, which are often bound with strands of mRNA. Ribosome numbers vary greatly according to environmentally driven growth conditions.

C. Biofilms

Until recently, perioperative professionals were taught that bacterial and yeast infections were caused by free-moving (planktonic), individual microorganisms or small isolated groups of microorganisms that entered the body via a wound or by direct invasion, multiplied and spread throughout the body, evaded immunological defenses (i.e., T-lymphocyte, B-lymphocyte, and phagocytic cells), and were shed from the body to infect new hosts [55].

Although it is true, in acute infections, that bacteria are generally found in a planktonic form, if bacteria establish a presence of any duration in the body, they tend to form a highly complex, self-regulating, bacterial community known as a *biofilm matrix* [56] that generally is both difficult and expensive to treat effectively. Terms relevant to biofilm bacteria are defined in Table 3.1.

Through a process termed *quorum-sensing*, bacteria in a biofilm matrix can chemically communicate system-level needs for the well-being of the entire biofilm community. Quorum-sensing between bacteria enables a biofilm community to induce or repress specific gene expressions regulating activities such as cell division, metabolic rates, production of virulence factors, plasmid transfer (often conferring antibiotic resistance), and release of planktonic bacteria from the biofilm [57, 58, 59].

Biofilm infections often originate due to contamination during surgical procedures, such as insertion of vascular catheter lines, pacemakers, heart valves, permanent biomaterials for repair of aneurysms, or prosthetic joint replacements. Other medical procedures associated with biofilm establishment are intratracheal intubation needed for ventilators and protracted use of indwelling urinary catheters [60]. It is important for perioperative staff members to recognize that implantation of devices or biomaterials may lead to the formation of biofilms, which will increase the risk of difficult-to-treat infections in postoperative patients.

1. Biofilm Genesis

To form a clinically significant biofilm, bacteria attach to tissue or an inanimate surface (e.g., titanium, stainless steel, polytetrafluoroethylene, Teflon®, polyester fiber) in a patient's body and then attract and attach to other planktonic bacterial cells [44]. Typically, direct attachment of bacteria to tissue elicits such a strong immunological response (e.g., high fever, malaise) that it becomes apparent, and patients are treated

TABLE 3.1

Definitions: Biofilm Bacteria Terms

Antimicrobial residual properties: When chlorhexidine gluconate or zinc pyrithione solution is used for at least 2 to 3 days before surgery as a presurgical site wash, the medications adsorb to the stratum corneum and prevent normal microbial population regrowth. Then, when the preoperative skin preparation is performed just before a surgical procedure, the normal microbial population numbers have already been reduced greatly, and the preoperative skin preparation is more effective.

Biofilm: A complex community of microorganisms enclosed in an exopolysaccharide matrix attached to tissue or an inanimate surface.

Coadhesion: The process of planktonic microorganisms binding to microorganisms attached to surfaces.

Exopolysaccharides: Polymerized material produced by microorganisms that constitutes the biofilm, providing protection and containment for the microorganisms.

Laminin: Linking proteins of basal lamina, which induce adhesion and enhance spreading of microorganisms in a biofilm.

Planktonic: Free-floating microorganisms within a biofilm.

Quorum-sensing: Chemical communication between microorganisms.

van der Waal's force: Nonspecific attraction between atoms that are 3 angstrom units (Å) to 4 Å apart.

---

immediately according to standard protocols before a biofilm is able to develop fully. In comparison, bacteria adhering to inanimate surfaces, such as implanted biomaterials and prosthetics, do not elicit an immune response, and patients are not treated for infection. Thus, normal skin residents (e.g., *Staphylococcus epidermidis*) attaching to implanted materials can lead to the development of a biofilm.

Bacterial attachment to inanimate surfaces generally requires that a surface be conditioned by autochthonous organic deposits, such as collagen, laminin, fibrin, and fibrinogen. The bacterial cells and organic deposits are mutually attracted via noncovalent forces, including van der Waal's forces and hydrophobic interactions [61, 62, 63]. Further, bacteria-bearing receptor sites for the organic compounds can attach directly by primary adhesion (Figure 3.13), divide, and begin producing an exopolysaccharide biofilm matrix. This establishes a bacterial presence with a much enhanced protection from the body's natural immunological surveillance, including phagocytosis, and from antibiotic treatments [61].

The conditioning layer influences which organisms will be the primary colonizers of the biofilm (Figure 3.14). For example, the organic substances noted above, when deposited on the inanimate surface, are conducive to biofilm formation by *Staphylococcus aureus*. For *Staphylococcus epidermidis* to produce a biofilm matrix, deposition of fibrinogen-binding protein is enhancive.

Various bacterial species that cannot attach to organic material themselves are able to attach to bacteria already adhering to the organic material (Figure 3.15). Specific compatible attachment sites are required among species in order to complete this process, termed *coadhesion*. The exact extent of this phenomenon and its process are not well understood at this time, but much anecdotal knowledge derives from oral biofilms (i.e., plaque), for which the coadhesive process is understood more fully [64, 65].

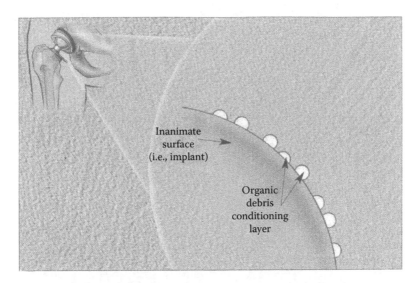

FIGURE 3.13 Surface conditioning. Organic debris "conditions" the inanimate surface of implants and other biomaterials.

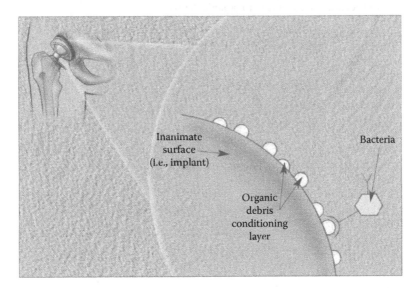

FIGURE 3.14 Bacterial attachment. Bacteria can attach to the organic debris by chemical adhesion.

Individual bacterial cells can form an elaborate matrix of exopolysaccharide and interstitial fluid consisting of 95 to 99% water, 2% bacterial content, and 1 to 2% exopolysaccharide content [56]. A biofilm matrix may appear as depicted in Figure 3.16 or as a thin layer of bacteria in an exopolysaccharide matrix. To date, no universal configuration has been determined for medical biofilms.

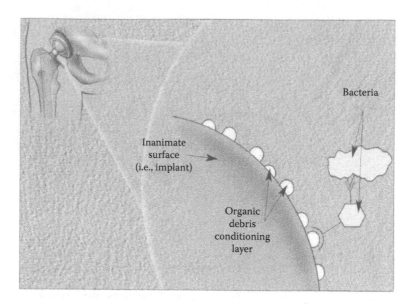

FIGURE 3.15 Coadhesion. Some bacteria that cannot attach directly to the organic debris can attach to bacteria that can attach to organic debris.

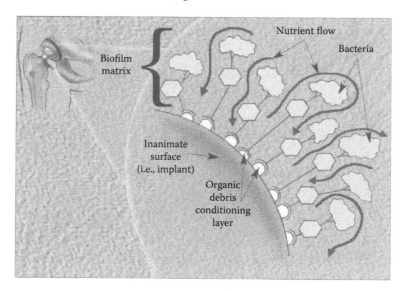

FIGURE 3.16 Biofilm matrix. The individual bacterial cells form an elaborate matrix of exopolysaccharides and interstitial fluid.

A biofilm matrix offers bacterial protection, and thereby increases resistance to the immunological responses of both humoral and cellular derivation, as well as from the phagocytic activities of neutrophils and tissue macrophages [56, 57, 58]. This enables growth of the biofilm matrix (e.g., on a venous catheter), with the sloughing off of planktonic bacteria that can produce septic conditions throughout a patient's body. Biofilm matrices often are slow growing and localized, however, affecting only an implant and surrounding tissues, and this may require surgical removal of the implant and debridement of associated tissues [66, 67, 68, 69].

Bacteria in a biofilm are 500 to 1,500 times more resistant to antibiotic therapy than are planktonic bacteria [70, 71]. Initially, researchers believed that the exopolysaccharide matrix provided a barrier that protected the bacteria from direct exposure to antibiotics [56, 57]. It now appears that the reason is more complex [70, 71]. Many of the bacteria in biofilm are metabolically quiescent, which limits their uptake of antibiotics. For the same reason, bacteria in biofilm generally do not replicate as rapidly as they would in the planktonic state, and different portions are in various stages of growth (i.e., static, stationary, exponential) at any given time [72, 73]. On average, however, the growth rate of the biofilm community appears overall quiescent [69, 70]. Because antibiotics inhibit synthesis of the bacterial cell wall or cell membrane, block protein synthesis at the 30 S or 50 S ribosome subunit, block DNA replication, or block folate coenzymes needed in DNA synthesis, bacteria in the biofilm most susceptible to antibiotics are in the exponential growth phase [72]. The antibiotics require an active metabolism and cellular division to be effective. However, destroying biofilm bacteria in the exponential growth phase leaves those in the static and stationary growth phases.

Resistance to some disinfectants (e.g., hydrogen peroxide) is related directly to bacterial density in a biofilm. Degradation of hydrogen peroxide by the catalase produced by bacterial cells, including nonviable cells, requires a concerted systems effort by a group of bacteria. A single bacterium is unable to produce enough catalase to overcome the debilitating effects of hydrogen peroxide [73, 74].

2. Perioperative Implications

Implanted devices (e.g., hemodialysis grafts, genitourinary prosthetics, pacemaker leads, prosthetic heart valves, vascular grafts) have significant potential for incurring biofilm infection [62]. For example, polyester fiber grafts that are used to replace and repair stenotic thoracic arteries and abdominal aortic aneurysms are prone to biofilm infections due to coagulase-negative *Staphylococcus* species and other species.

3. Vascular Catheterization

Biofilms are a serious concern for patients who have vascular catheters. Microorganisms, particularly normal skin flora, colonize and form biofilms quickly on catheter surfaces; however, contaminative exogenous microorganisms from healthcare personnel, contaminated infusion fluid, and distal infections transported via hematogenous routes also have been implicated [61, 66, 75]. Many of the millions of patients who undergo vascular catheterization procedures in the United States every year suffer from bacterial biofilm infections [70]. Of patients who have a central venous catheter, 4 to 14% experience septicemia [69], often due to normal skin bacteria, such as *S. epidermidis*. Other common bacteria cultured from vascular catheters include *S. aureus*, *Pseudomonas aeruginosa*, and *Enterococcus* species [75].

Local biofilm infections, very common at catheter-insertion sites, include tunnel infections (cellulitis along the subcutaneous catheter route) and frank catheter-tip colonization. These can lead to life-threatening septicemias.

The current trend for topical antimicrobials is to demonstrate antimicrobial persistence for longer periods of time to limit injury to veins from multiple removal and insertion procedures. However, patients with indwelling catheters have an increased risk of acquiring biofilm infections when the catheter remains at a specific site for a prolonged period. To counteract this threat, some catheter manufacturers are partnering with manufacturers of topical antimicrobials to produce tubing, cannulas, and catheter insertion tips treated with antimicrobial products, such as silver sulfadiazine and chlorhexidine gluconate, and/or antibiotics, such as minocycline and rifampin [76].

Nearly 100% of superior vena cava catheters become locally infected within 2 to 3 days of placement [70], often with coagulase-negative *Staphylococcus* species, especially *S. epidermidis*. A high proportion of these bacteria demonstrate resistance to multiple antibiotics, especially to the penicillins [70].

Devices exposed to direct blood flow, such as vascular catheters and heart valves, pose a serious risk of biofilms and chronic systemic infection. It is important that perioperative nurses understand the implications of Virchow's triad (surface area–blood contact–flow rate), and that the greater the surface area, the more probable it is that bacteria will colonize it. Direct blood flow offers a continuous source of conditioning material that prepares a device for bacterial attachment, and subsequently provides a shearing effect that can transport planktonic bacteria and biofilm clumps to other areas of the body [62, 77]. Finally, ventricular-peritoneal shunts for reducing intracranial pressure almost always become contaminated with *S. epidermidis* and *S. aureus* biofilms [69].

4. Orthopedics

Joint replacements (e.g., hip, knee) carry a threat of postoperative biofilm infections having particularly devastating effects, including osteomyelitis. *S. epidermidis*, *S. aureus*, and *Pseudomonas aeruginosa* are commonly cultured from these implants [67]. In many situations, biofilm-infected joints require revisional surgery with a second implant, and this can be expensive and traumatic. Special precautions with topical skin antiseptics may be valuable, because *Staphylococcus epidermidis* is prevalent in these infections.

5. Endotracheal Tubes

Patients who remain intubated after a surgical procedure are prone to ventilator-associated pneumonia. Endotracheal tubes bypass the body's normal pulmonary clearing responses (e.g., coughing, mucociliary clearance), increase mucus secretions because of irritation and inflammation, and tend to denude cilia from the tracheal epithelium. This allows secretions to enter the lungs through the stented glottis [62]. Biofilms quickly develop on the endotracheal tube and can pass easily into the lungs, particularly during suctioning procedures.

D. Are Topical Antimicrobials Effective?

It is pertinent to determine the effectiveness of topical antimicrobials that are used to remove germs from the skin before catheter insertions or before preoperative skin preparation or when challenged with bacteria in a biofilm matrix. Currently, antimicrobial efficacy testing is performed almost exclusively on bacteria in the planktonic state. An evaluation to determine efficacy against biofilms was performed using a number of common topical skin antiseptics to challenge pathogenic bacterial species prevalent in biofilm infections. The evaluation determined the resistance to killing provided by a biofilm matrix compared to the bactericidal effectiveness of each antiseptic versus the bacteria in a planktonic state.

III. MYCOLOGY (FUNGI)

Fungi, unlike bacteria, are eukaryotic cells that contain at least one nucleus, nuclear membrane, endoplasmic reticulum, and mitochondria [44, 78]. Fungal cells are much larger than bacteria and are closer structurally and metabolically to cells of higher plants and animals [79]. Most fungi possess a rigid cell wall and some species produce motile, flagellated cells. Unlike most members of the plant kingdom, fungi are nonphotosynthetic [80].

Fungi may exist as single oval cells (yeasts) that reproduce by budding, or as long tubular strands (hyphae), usually septate, which exhibit apical growth and true lateral branching [33, 44, 78]. Branching intermingled and often fused, overlapping hyphae constitutes the mycelium that forms the visible mold colony [79, 80]. Reproduction may be vegetative (asexual) via variously specialized germinal cells called *conidia*, or it may consist of elaborate specialized structures that facilitate fertilization, protection, and dissemination of the resultant spore. Conidia may be simple fragments of a hypha, or they may be produced from variously specialized structures called *coniophores*. Spore-bearing structures may be simple lateral branches of the hyphae or they may be constructed into large reproductive bodies such as mushrooms and bracket fungi that protect the spores as they develop and facilitate their dispersal at maturity.

The natural habitats of most fungi are water, soil, and decaying organic debris, and most are obligate or facultative aerobes [79, 81]. They are chemotropic organisms, many of which secrete enzymes that degrade a wide variety of organic substrates into soluble nutrients that can be absorbed.

A. YEASTS

Yeasts are single cells, usually spherical to ellipsoidal, ranging from 3 to 15 µm in diameter [32, 78]. Most reproduce by budding, but a few do reproduce by binary fission.

After growth on agar media (24 to 72 hours), yeasts tend to produce colonies that are pasty and opaque-looking, generally growing in colonies from 0.5 to 3.0 mm in diameter. A few species have characteristic pigments, but most are cream colored. Using microscopic and colonial morphology, it is difficult to distinguish species, so nutritional studies and specialized media must be used [79, 80].

B. MOLDS

The molds are growth forms that are multicellular and filamentous and appear in colonies [59, 60]. The colonies consist of branching cylindrical tubules varying in diameter (2 to 10 μm) termed *hyphae*, which grow by apical elongation. Hyphal tips contain densely packed membrane-enclosed vesicles, many of which fuse with the cell membrane during active growth [44]. The mass of intertwined hyphae that accumulates and constitutes the visible mold is termed the *mycelium*. The hyphae of some species are divided into cells by cross-walling (septa) that form at regular, repeating intervals during filamentous growth and, in some, the septa are penetrated by pores that permit flowing of cytoplasm but not organelles.

Molds are extreme opportunists, tending to grow well on a tremendous variety of substrates found in the environment and on most laboratory culture media [79, 81]. Hyphae that actually penetrate into the substrate/media are termed *vegetative of substratic hyphae* and serve to anchor the mycelium. Those hyphae that extend above the surface (aerial hyphae) usually bear the specialized structures (conidiophores) that produce the asexual reproductive cells, the conidia.

C. DIMORPHISM

Most fungi exist only as a mold or a yeast form, but a number of species, including several important pathogens in humans, are capable of growing as both a yeast and a mold [32, 44, 78]. Temperature plays a major role in this dimorphism. At about 35 to 37°C, such species grow in the yeast phase, but at lower temperatures (20 to 30°C), they grow as molds. Although this is referred to as *thermal dimorphism*, available nutrients, carbon dioxide levels, cell density, age of culture, or a combination of these factors can also induce a shift to morphism.

1. Cell Structure

Fungal cells are composed of a cell wall, cell membrane, and cytoplasm containing an endoplasmic reticulum, nuclei, nucleoid, storage vacuoles, mitochondria, and other organelles (Figure 3.17).

2. Capsule

Some fungi secrete an external slime layer [78, 81]. This slime layer, or capsule, is composed mostly of amphorous polysaccharides that play a major role in the cell's structure. The capsular composition varies among species in quantity, chemical composition, antigenic properties, and physical attributes such as solubility and viscosity.

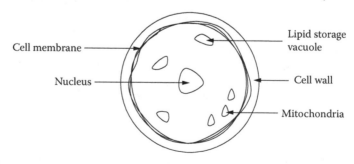

FIGURE 3.17   Basic fungal structure.

The capsule component does not affect the permeability of the cell wall and cell membrane, but due to its slime nature, it probably influences growth by preventing the disassociation of buds from yeast cells and the dispersion of yeast in air and water. As in bacteria, the fungal capsule is thought to be associated with virulence.

3. Cell Wall

The cell wall is composed of 15 to 30% of the dry weight of a fungal cell [32, 44, 46]. It provides rigidity and strength to the cell as well as protection from osmotic shock. As in bacteria, the cell wall determines the morphological shape of the fungi. The cell wall is, on the average, thicker in yeasts than in the hyphae molds.

About 80% of the cell wall's composition is carbohydrate material [44, 78]. Several different types of polysaccharides are found and vary in specific type and abundance from species to species. The major fungal cell polysaccharide components include chitin, chitosan, cellulose, (3-glucan, a-glucan, and mannan) [32, 44, 52]. Because the polysaccharides shared by fungi are common, many fungi groups exhibit the same type of surface antigens. However, the side chains of the glucans vary considerably in the number, length, and linkage of their residues, and present many unique antigens. There are many common antigens unique to species, as well as those shared in common with other species.

About 10% of the fungal cell wall is made of protein and glycoprotein [32, 52]. Proteins include enzymes involved in the cell wall growth, specifically extracellular enzymes and structural proteins that provide cross-linkage to the polysaccharide chains. The protein concentration increases at the inner cell wall surface. Cell wall proteins contain much sulfur in their amino acids, which is linked by disulfide bonds. However, these are more common in walls of hyphae than in yeast walls. The disulfide bonds appear to play a role in the transformation from mold stage to yeast stage of growth, among the dimorphic species.

Cell wall polysaccharides are fibrillar and multilayered in appearance and look like long microfibrils microscopically [79, 80]. Four to eight distinct cell wall layers are usually observed, but the degree of organization is variable. Generally, the most protein-laden layer is nearest to the cell membrane [32, 44, 79]. The external layers tend to be less well organized and less compact. Some fungal cell walls contain tightly interwoven microfibrils embedded in an amphorous, polysaccharide matrix.

In a fungal infection, the host is initially exposed to the cell's surface, and research suggests that the cell wall governs the pathogenicity of the fungus [78, 81]. The cell walls of the major pathogenic fungi possess components that mediate host cell attachment (e.g., phagocytes, epithelial cells, and endothelial cells). Cell surface ligands, receptors, and other cellular components promote colonization and invasion of host cells and resistance to host immunological defenses [82].

Host immune response to fungal cell wall antigens is very strong and quite specific [32, 44, 82].

4 Specific Microorganisms

I. STAPHYLOCOCCUS

Staphylococcus epidermidis and *S. aureus* are two very common and clinically significant bacteria in hand colonization.

S. epidermidis normally colonizes skin surfaces without causing the host any ill effects. However, if these microorganisms gain entry, through broken skin for example, they can become pathogenic and can cause much damage to a person's system. When the skin is incised during a surgical procedure, these microorganisms can enter the sterile body and cause infection. Once they pass the skin barrier, these microorganisms can invade deeper tissues. *S. aureus* are not generally pathogenic to humans, but patients who have serious underlying diseases or who have undergone surgical or other invasive procedures are at greatest risk for infection [32]. *S. aureus* is the most virulent of these two microorganisms, and they can produce biofilms.

S. epidermidis produces a slime or biofilm to enhance microorganism adhesion and provide refractory conditions to antibiotic action with a propensity to acquire antimicrobial resistance.

S. aureus produces and secretes toxins and enzymes that have a role in its virulence. It can produce alpha, beta, gamma, and delta toxins to act on host cell membranes and mediate cell damage. Leukocidin tends to aid in the destruction of phagocytes, monocytes, and basophils.

Also of particular interest is the community-associated infection, such as methicillin-resistant *S. aureus* (MRSA) and community-acquired MRSA, or CA-MRSA.

II. STREPTOCOCCUS

Despite the current availability of antimicrobial agents, *Streptococcus pneumoniae* remains a leading cause of morbidity and mortality not just in older adults, but through the entire population. It is the most common cause of bacterial pneumonia, as well being an important cause of meningitis, bacteremia, and septicemia.

S. pneumonia is an encapsulated Gram-positive coccus, oval or spherical in shape, and 0.5 to 1.25 μm in diameter. *S. pneumonia* is lancet-shaped, and as seen in direct culture of sputum and body fluids, has been observed individually, in pairs, and in short chains. The microbe is very sensitive to products in its fermentative metabolism capsule, which may be readily observed in wet mounts of virulent organisms using India ink.

A. ANTIGENIC STRUCTURE

Capsular antigens that consist of complex carbohydrates form hydrophilic gels on the surface of the microorganisms. The polysaccharides are antigenic and are the basis for the separation into 84 different subgroups [32].

Somatic antigens. C-polysaccharide is a species-specific carbohydrate that is a major structural component of the cell wall of all pneumococci. It is a teichoic acid polymer containing phospholine as a major antigenic determinant. It is responsible for the agglutination of pneumococci by certain specific myeloma proteins and the interaction of the polysaccharide C with Serum B-globulin in the presence of calcium [32].

F Antigen. Another major antigenic component of the pneumococcus is the Forssman or F antigen. The F antigen is a lipoteichoic acid and contains choline as a constituent of teichoic acid. The F antigen is a strong inhibitor of peptidoglycan hydrolase in a human [32].

M Protein. Type-specific protein antigens analogous to the M protein of *Streptococcus pyogenes* yet immunologically distinct are present in pneumococcus. No real correlation has been demonstrated between the presence of a specific type of M protein and the type of organism based on capsular polysaccharide antibodies. M protein do not inhibit phagocytosis, and are therefore not proteins [32].

III. OTHER MICROORGANISMS

A. MYCOBACTERIA

The most distinctive property of mycobacterium is their characteristic staining. They are difficult to decolorize with acid alcohol. Therefore, they are referred to as acid-fast bacilli [32].

Within the genus are species responsible for two dreaded diseases—tuberculosis and leprosy. *Mycobacterium tuberculosis* is usually considered a complex of *M. tuberculosis*, *M. bovis*, *M. bovis* BCG, and *M. africanum* [32]. All these species are capable of causing tuberculosis. It is estimated that 1.7 billion people or one third of developing countries are infected with this complex.

B. ENTEROBACTERIACEAE

Historically, the enterobacteriaceae have been divided into opportunistic and intestinal pathogens [32]. The intestinal pathogens include *Salmonella*, *Shigella*, and *Yersinia*. The opportunistic bacteria have been included in other genera. Recent developments in the genetic relationships of *Escherichia coli* and *Shigella* have made this distinction less clearly defined.

C. ESCHERICHIA COLI

The microorganism is Gram-negative, a facultative inhabitant of the large intestine. It is commonly isolated in urinary infections, wound infections, pneumonia, and septicemia.

E. coli grows well in commonly used media and most strains are lactose fermenting [32]. Those strains particularly associated with urinary tract infections are B-hemolytic on blood agar. The majority of isolates are nonpigmented and mobile.

1. Surface Antigens

The K1 capsule is unique among capsular antigens of *E. coli* [32]. It enables the bacteria to resist killing by neutrophils and normal blood serum. The antigen is frequently found in patients with bacteremia and meningitis.

E. coli produces a number of different types of pili that enable the microorganism to attach to various tissues successfully. For example, in urinary infections, they are not removed by urinary flow. The precise role of other surface antigens is not known [32]. Certain serotypes have been associated with different diseases; whether the genetic location of pathogenic determinants is merely due to the closeness of surface antigens has not been determined.

2. Enterotoxins

E. coli plays a significant role in gastrointestinal infections. The ability of *E. coli* to possess the necessary plasmid (LT) is similar to the enterotoxin of other microorganisms (e.g., *Vibrio cholerae*) [32].

E. coli produces at least two types of toxins termed *verotoxins*. Verotoxin-producing *E. coli* (VTEC) have been associated with three syndromes: (1) diarrhea, (2) hemorrhage, and (3) hemolytic uremia syndrome (HUS). Because of these similarities to Shiga toxins, these toxins have been referred to as Shiga-like toxins (SLT) [32].

D. KLEBSIELLA

The most commonly isolated member of the genus *Klebsiella* is *K. pneumoniae*, a causative agent of pneumoniae [32]. It is found in patients, especially those who are hospitalized and have urinary catheterization. It is primarily a strong opportunistic microorganism, so it can infect other sites besides the urinary system, in particular, the respiratory tract.

There are six species to which humans are susceptible, but we will limit our discussion to *K. pneumoniae*. It is a Gram-negative lactose-fermented species that is nonmotile. The capsule enables the microorganism to resist phagocytosis [32].

The microorganisms are responsible for a wide range of infections, including pneumonia, urinary tract infections, bacteremia, and meningitis.

E. SERRATIA

Serratia marcescens, once thought to be a harmless microorganism, turns out to be a major producer of nosocomial infections [32]. However, this microorganism is used in the current Food and Drug Administration (FDA) guidelines for evaluating healthcare personnel handwash products. Given it is tested according to the guidelines, it remains safe.

S. marcescens has the advantage that certain growth media distinguish these colonies from *Staphylococcus epidermidis* by staining the media red, instead of a white color as by *Staphylococcus*, which makes this microorganism (contaminated) critical for determining a bacteria-covering *Staphylococcus epidermidis* (a resident bacteria).

F. *PROTEUS BACTERIUM*

Two Gram-negative species concern us in this work: *P. mirabilis* and *P. vulgaris*.

Both of these microorganisms are actively motile at 37°C and produce translucent sheet growth on nonselective agar, such as blood agar, causing a swarming phenomenon [32].

G. *SALMONELLA*

The *Salmonella* species is complicated by the development and utilization of several different nomenclatures over the years. For example, the Kauffman-White antigenic scheme gave rise to over 2,000 species. This was wrong. Ewing and coworkers [32] proposed that there are only three species: (1) *S. cholerae*, (2) *S. enteritidis*, and (3) *S. typhi*. All other species were serotypes of *S. enteritidis*. The three species designated were used by the National Salmonella Center from 1972 to 1983 but were not in use by other countries. Since that time, genetic studies have revealed that all *Salmonella* and microorganisms that were classified in the genus *Arizona* belong to the same species in a genetic sense. Differences in antigenic types were caused by a divergence within a single species, *S. enterica* [32].

To correct this, the species has been redesignated into five subgroups based on DNA. These now are (1) Salmonella Subgroup 1, with subspeciation *enterica*; (2) Salmonella Subgroup 2, with subspeciation *salamae*; (3) Salmonella Subgroups 3a and 3b, with subspeciation *arizonae* and *diarizonae*; (4) Salmonella Subgroup 4, with subspeciation *houtenae*; and (5) Salmonella Subgroup 5, with subspeciation *bongori*. Although any human infection can be caused by any of the five groups, the major pathogen is found in Subgroup 1.

H. *SHIGELLA*

This organism is a major cause of bacillary dysentery. Most disease is seen in pediatric age groups with the majority occurring in children 1 to 10 years old. In the United States, 15% of all pediatric cases of diarrhea have been estimated to be from *Shigella* [32].

The *Shigella* microorganism appears indistinguishable from *Escherichia coli*. Because the majority of *Shigella* strains cause dysentery, but the majority of *E. coli* do not, they appear to be different. *Shigella* is divided into four major subgroups: (1) Serogroup A, *S. dysenteriae*; (2) Serogroup B, *S. flexneri*; (3) Serogroup C, *S. boydii*; and (4) Serogroup D, *S. sonnei* [32].

Serogroups A, B, and C are biologically similar, while Serogroup D is distinct. They are nonmotile and do not produce H_2S (except *S. flexneri* [Serogroup B]) from glucose [32].

1. Toxins

Shigella species carry the protein toxin, which, depending on species, produces great amounts of toxin, and therefore more severe diseases [32]. The toxin has a multiplicity of effects, including neurotoxicity, cytotoxicity, and enterotoxicity effects.

IV. FUNGI

All fungi are eukaryotes, instead of prokaryotes, which is a difference among these bacteria [32]. All fungi are Gram-positive, so the staining effect is of little practical concern as it does not help in their identification.

Fungi grow dimorphically in two basic forms: yeasts and molds. Yeasts refer to unicellular growth of fungi. They are usually spherical, varying from 3 to 15 µm. Molds refer to multicellular filamentous growth. They consist of colonies of branching cylindrical tubules, termed *hyphae*. Hyphae types are densely packed with membrane-enclosed vesicles.

A. DIMORPHISM

Some species of fungi are dimorphic, or they can grow as yeasts at 37°C, or as molds at colder temperatures, such as 25°C.

For example, yeast such as *Candida* can grow in a person's mouth or vagina. A fungi can grow as a mold on skin surfaces, such as on the toenails.

B. CANDIDA

These organisms are members of the normal flora of the skin, mucous membranes, and gastrointestinal tract [32]. Of more than 100 species, *Candida* species are important human pathogens that pose an ever-present risk of infection, particularly opportunist infections, especially in immunocompromised hosts (e.g., transplant patients, acquired immunodeficiency syndrome [AIDS] sufferers, cancer patients).

C. ASPERGILLUS

Among 150 different species and subspecies of recognized *Aspergillus*, many have been known to produce infection [32]. Aerosolized *Aspergillus* spores are found nearly everywhere, so individuals are routinely exposed to them. *Aspergillus* can and cause disease in three major ways: (1) through the production of mycotoxins, (2) through induction of allergenic responses, and (3) through localized or systemic infections. The most common pathogenic species are *A. fumigatus* and *A. flavus*.

D. HISTOPLASMA

Histoplasma is among the pathogenic fungi. *H. capsulatum* can cause disease in humans, dogs, and cats. It is endemic in certain areas of the United States, and infection is most commonly due to inhaling contaminated air containing these fungal cells [63].

1. Cytoplasmic Membrane

Fungi have a bilayered, cytoplasmic membrane similar to that of other, higher eukaryote cells [52, 78]. The membrane protects the cytoplasm, regulates the intake and secretions of solutes, and assists in the synthesis of the cell wall [32]. The membrane contains several phospholipid types, their type depending upon the species of fungus. The most common include phosphatidylcholine and phosphatidylethanolamine, but phosphatidylserine, phosphatidylinositol, and phosphatidylglycerol are also found in lesser degrees [52]. The total phospholipid content of the cytoplasmic membrane varies between species, strains of a species, and within those strains, depending on environmental conditions.

2. Cytoplasmic Content

Often, multiple nuclei are seen in both yeast and mold cells [45, 47]. For example, when septa are not present in mold hyphae (the zygomycetes), there is a plasmic continuity, and the hyphae are considered multinuclear. Among the "higher" fungi, the septate hyphae often possess a central pore in the septum that allows mixing of cytoplasmic content. Many species are capable of dilation/constriction of the septal pores, allowing the flow of organelles, including nuclei, mitochondria, lipid vacuoles, and ribosomes, between the hyphal cells.

Fungal mitochondria resemble those of both plant and animal eukaryotic cells [32]. The number of mitochondria within each cell can vary but are directly related to the level of cellular respiration. During sporulation, for example, as the cell requires increased energy, the number of mitochondria increases within each cell. In tissues, fungi tend to have fewer mitochondria [60].

The cells of many fungal species contain vacuoles that function as complex organelles. Some vacuoles contain a variety of hydrolytic enzymes, while others serve as storage depots for ions and metabolites such as amino acids, polyphosphates, and other compounds.

V. VIRUSES

Viruses are a unique life form quite distinct from both the prokaryotic or eukaryotic organisms [32, 44]. They are much smaller than bacteria and fungi and are obligate parasites that must live within the cells of host organisms. They consist of a single type of nucleic acid—DNA or RNA, either single- or double-stranded—encased by a shell (capsid) structured primarily of protein. The capsid of some viruses is further enclosed by a unit membrane consisting of lipids and glycoproteins (so-called *spike proteins*) [32, 50, 52]. Viruses do not reproduce by binary fission because they lack the intracellular components necessary to produce macromolecules. Instead, viruses capture and redirect the nuclear organelles of a host cell to synthesize new virus components, which then autoassemble within the host cell into *virions*, which are complete viral particles [32, 44, 48, 49].

Virions range in size from 20 to 25 mm. The viral nucleic acid is enclosed by a capsid that generally is either a helix or an icosahedron. Some viruses, however, take neither form, and so are termed *complex viruses*. The nucleic acid/capsid is referred to as the *nucleocapsid*. The capsid itself is composed of single polypeptide chains, termed *structural units*, which may combine and form the multiple polypeptide units termed *capsomeres*, or they can combine directly to form the helix-form capsids [49, 82]. The capsomeres tend to form the more complex icosahedral capsids. The mature infectious virus may consist of the nucleocapsid (capsid containing nucleic acid) alone, or may be surrounded by an envelope consisting of lipids and glycoproteins. It is acquired during the process of *budding* through the cytoplasmic membrane of the host cell [47, 82].

Viruses are generally grouped into five morphological categories (Figures 4.1 to 4.4).

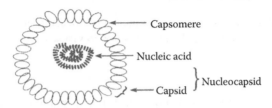

FIGURE 4.1   Naked icosahedral virus.

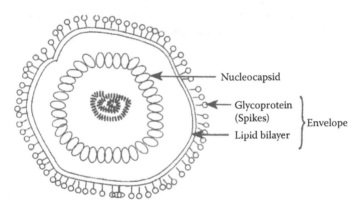

FIGURE 4.2   Enveloped icosahedral virus.

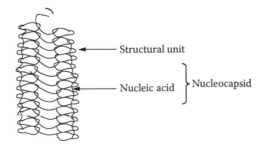

FIGURE 4.3   Naked helical virus.

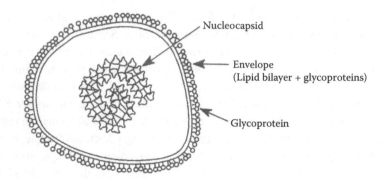

FIGURE 4.4   Enveloped helical virus.

However, nonenveloped viruses are much more resistant to alcohols, chlorhexidine gluconates, and iodophors than are enveloped viruses.

A.   COMPLEX VIRUSES

The fifth category includes a number of viruses that do not fit into any of the previous four groups and are termed *complex viruses*. The poxviruses provide examples of the complex structural form [50].

We will not spend much time on viruses, because the FDA does not currently have any tests for claims that one must achieve.

5 Measurement of Antimicrobial Action of Topical Antimicrobials

Before going into detailed discussion of evaluative procedures, let us look at the general performance characteristics we intend to evaluate.

I. GENERAL PERFORMANCE CHARACTERISTICS

Generally, topical antimicrobials are evaluated for their antimicrobial efficacy in terms of three parameters: (1) the immediate degerming effect of the formulation, (2) the persistent antimicrobial effect of the formulation, and (3) the residual antimicrobial effect of the formulation, when appropriate [14, 84, 85, 86].

A product's immediate antimicrobial effectiveness is a quantitative measurement of both the mechanical removal of microorganisms by washing and the product's ability to rapidly inactivate microorganisms residing on or contaminating the skin surface [14]. The persistent antimicrobial effectiveness is a quantitative measurement of a product's ability to prevent microbial recolonization of the skin surfaces, either by microbial inhibition or lethality after application of the product. The residual effectiveness is a measurement of the product's cumulative antimicrobial properties after it has been used repeatedly over time. That is, as the product is used repeatedly, it is absorbed into the stratum corneum of the skin and, as a result, prevents microorganism recolonization of the skin surfaces.

The ability to measure these three parameters poses certain problems to the investigator [14]. Because a product's antimicrobial activity is relative, it is of utmost importance that its efficacy measurements be well defined and clearly stated before conducting an evaluation [30]. Let us turn our attention to some additional, important points to consider when evaluating topical antimicrobial products.

Traditionally, topical antimicrobial evaluations have been, logically and appropriately, the domain of the microbiologist, since evaluations are in terms of microbiological parameters [14]. Although the focus is on microbiology, the efficacy of the product is measured mathematically, specifically through biostatistical methods. Hence, because of the bipartite structure necessary in these evaluations, ideally, the investigator has a strong background in clinical microbiology as well as an experimental design capability that will unambiguously answer the questions of interest.

Because it is the experimental design and the statistical evaluation that customarily give investigators the most trouble, we will concentrate mainly in these areas. Our general strategy, then, will be to design, conduct, and evaluate the study by using experimental design concepts, microbiology, and statistics, as appropriate [2, 14].

A statistically designed study is one that systematically collects, organizes, analyzes, and draws valid conclusions about the antimicrobial products evaluated [87]. When one designs a clinical trial, it is important that the research objectives be explicitly stated and that the study be designed to achieve those objectives clearly, concisely, and unambiguously. This is the key to a successful evaluation program, and it has been our experience that this approach appreciably expedites Food and Drug Administration (FDA) review [14]. Now, let us look at how a study is designed based on clearly defined study objectives.

Simply stated, the study objectives are to provide answers to questions the study is intended to address. These will be perhaps the easiest to draft of all the necessary documentation relevant to clinical trials, but they are among the most important. Every other document or concern of the study will be in support of the objectives. Given the objectives, the general schema will be to design the study relative to those objectives to conduct the study employing appropriate microbiological methods, and finally, to evaluate the results applying predetermined, appropriate biostatistical methods. To accomplish this efficiently, several conceptual components should be clearly conceived. These include descriptions of the study's purpose and the experimental strategy to achieve the objectives.

A.   Description of the Study's Purpose

A concise description of the study's guiding purpose is important [88]. Currently, this is written as the primary and secondary purpose of the study. This point is so implicitly obvious that it is often explicitly ignored until the entire study has been completed, by which time the original objectives have become obscure and unclear. Then, the investigator must backtrack through the study to determine what the original objectives were, and all too often, attempt to make the data relevant to those objectives.

In determining the study's primary purpose, the three parameters of antimicrobial efficacy measurement (the immediate, persistent, and residual effects) must be specifically considered [15]. If one is evaluating a prevenipuncture antimicrobial product, the primary purpose will likely address the immediate antimicrobial effects only. However, if one is evaluating a surgical scrub product, the immediate, persistent, and residual antimicrobial effects will be of importance. Also, importantly, there are four things that the researcher must check prior to believing the study results. One constantly hears about biasing the study by telling subjects too much or too little. This occurs, but it is subjective belief in what may happen. The four areas I will discuss view the study in objective—scientific—terms. The thing to do is to get the baseline data for a study and check them prior to evaluating the study. This is accomplished through the following four steps:

1. Perform exploratory data analysis (EDA) on the baseline data.
2. Check the standard deviation for any changes from past studies.
3. Ensure the study was properly randomized.
4. Check the control product data.

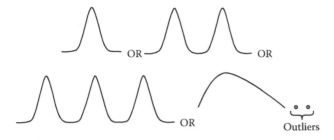

FIGURE 5.1 Possible data distributions.

1. Exploratory Data Analysis

First, perform a combined exploratory data analysis on the baseline data of all groups in the study. Specifically, create stem-and-leaf and letter-value displays to see if the data look biased. Generate these data displays using all the subjects who participated in the study or including only the subjects who achieved the required log_{10} counts. Determine what the differences would be. If all the data for subjects who passed the three log_{10} requirements are included, what do the data look like? Is there a normal distribution, or are there two or three distributions, or is it a skewed distribution with outliers (Figure 5.1)?

2. Standard Deviation

Is the standard deviation, s, the same as it has been in pilot studies for former studies? If it is larger—substantially larger—there is a problem. What about variation in standard deviations? Are they the same or are they different? Based on experimentation, I recommend using the Levene test to determine if they are the same or different [89].

3. Proper Randomization

Were the study participants sufficiently randomized? It must be apparent, exactly, if and how this was done. If one-quarter of the subjects were noncompliant with study requirements, where would they have been placed—test product, control product, or both products? If this has not been done, the study is invalid for the FDA.

4. Control Product(s)

Look at the control group of the study. Is it any different than previous controls on other studies of this type? An appropriate reference (control) product must be used in the study as a gauge for comparison of the total antimicrobial effects and to provide an internal validation of the study. If the control product responds as it has in previous studies ($\bar{x}, s$), the project looks good.

B. DESCRIPTION OF THE EXPERIMENTAL STRATEGY

Once the study's primary and secondary purposes have been determined, one can begin designing the study to meet those requirements [14]. The design of the study

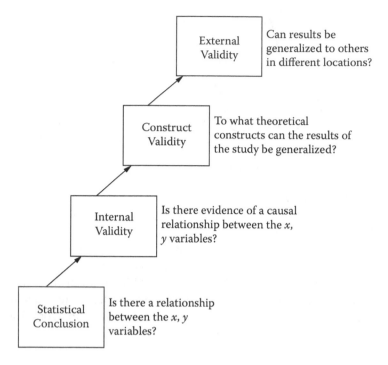

FIGURE 5.2   Prestudy parameters.

must be such that the study's objectives and purpose are achieved and based on valid results. Validity in experimental design is concerned with two areas: internal and external validity [30].

Before beginning, let us look at an example. Figure 5.2 displays the things that need to be known prior to conducting a study.

1.   Statistical Conclusion Validity

This concerns the inappropriate use of statistical procedures for analyzing the data, leading to invalid conclusions about the relationships between the independent variable (x) and the dependent variable (y).

The major concern that needs to be addressed, for example, with analysis of variance (ANOVA) models, is the s^2 (*mean square error* term). This value should be as small as possible. For example, if a statistic is to compare a preoperative skin preparation study using three products and four time frames, the model is

$$\hat{y}_{\log_{10}} = \text{Blocks} + \text{Times} + \text{Products} + (\text{Times} \times \text{Products}) + e ,$$

where $\hat{y}_{\log_{10}}$ = the dependent variable linearized with a $\log_{10}$ transformation and Blocks = a matched, paired t-test, utilizing a randomized block design. That is, each subject in this study receives two of the three products (one on each of their two sides).

The three products are A, B, and C, and the randomization are as follows:

		Randomization	
		Side	
Block	Subject	Left	Right
1	1	A	B
1	2	A	C
1	3	B	C

The blocked design means that there will be three subjects per block of two different sides of the body. To keep the subject size balanced, the sample size is 3, 6, 9, 12, 15, ... 60, 63 ... For example, the sample size for the study must be 60 or 63 subjects; it cannot be 61 or 62.

Times

0 = Baseline*
1 = 10 minutes
2 = 2 hours
3 = 6 hours

The easiest method to employ, no matter what the data look like, is to transform them into reduction data:

Reductions = Baseline – Wash Times (10 min, 2 h, or 6 h)
Products: 1, if Test Product; 2, if Positive Control, and 3, if Negative Control
Times × Product Interaction
e = Error term

2. Internal Validity

Is there actual evidence that x (the independent variable) caused y (the dependent variable)? For example, a researcher performs a study and finds that the product does not kill the necessary level of microorganisms. Is this correct?

The internal validity of a study can be assured by using the appropriate statistical experimental design [13]. But internal validity can also be built into the study by random sampling, by appropriate blocking, and by the use of specific statistical and experimental controls, as appropriate [90]. For example, the experiment might have caused the problem by having too much salt in the media. If the control also did not kill many microorganisms, this would be apparent.

There are various errors that can cause a study to be internally invalid. Two of them are routinely encountered in clinical trials [91]. They are as follows:

---

* An effort is made to know that the baseline values among the three products are the same. If it is not, a covariance model may be employed. The problem with covariance models is that, if the control product is a negative control, its slope will probably not be similar to the test product, and therefore, the test will be invalid. (Covariance models must have the same slope.)

1. *History* is the specific events occurring between the first and subsequent measurements. An example in clinical trials might be the subjects washing their skin surfaces with other antimicrobial compounds between the baseline and posttreatment measurements, causing the posttreatment measurement to be biased by the uncontrollable effects of extraneous antimicrobial compounds. The antimicrobial product being evaluated would score higher in efficacy than it actually is.
2. *Instrumentation* changes in the observations or measuring instruments (e.g., agar plates) may produce changes in the study results [14, 91]. An example in clinical trials could be different media lots, which are significantly different in nutritional characteristics and, therefore, extremely variable in their ability to support microbial growth. Another example is studies in which subjective evaluations are provided, like those measuring healthcare personnel skin irritation after use of various hand-cleansing products repeatedly over time. In this case, two different observers may score a degree of chafing or redness differently that is, in fact, the same.

3. Construct Validity

If the study passes the standard evaluation, then what constructs can be generalized from the results? This aspect must look at how the study was constructed and controlled—for example, if the subjects

1. had at least 1.0×10^5 microorganisms on the test site,
2. were between the ages of 18 and 65 years old, and
3. did not bathe for 48 hours prior to the test.

Then these are the conditions or constructs of the study.

If the test passed the requirements, then these results would be dependent upon the following:

1. Subjects having fewer than 1.0×10^5 microorganisms at the test site. What happens to people who have fewer? This cannot be attained by the study.
2. Subjects 17 years or younger and 66 years or older were not included in this study. What would happen if they used the product? You do not know.
3. If you state that *all* individuals, no matter how old, would benefit from using the product when they were not tested, you would be stating a fact that could not be supported.
4. Finally, a person who bathes before using this product would not be using the product correctly, according to the conditions of this study, testing was to occur 48 hours after bathing.

While this does not seem to make sense, it really does. Because the FDA requires a $3.0 \log_{10}$ reduction, these things [1, 2] must be addressed in the study to assure that microbial colony counts that normally reside on the skin are high. But this changes the outcome of the study. The second factor, age, is put in for safety reasons, but one can say nothing about the young or the old with regard to the product.

This brings us to a problem: undercontrol versus overcontrol. If one controls every aspect of a study very tightly, no doubt the study will have little variability. But this overcontrol would actually be valid only for those types of subjects who receive that treatment, under those specific conditions.

For example, subjects who had 1×10^5 microorganisms, who did not shower or bathe, who did not use any cosmetic products, who did not sunbathe, etc., would be included in the outcome. When the product is actually used by people, they would not be using the product as it had been tested, because the actual use of the products and the tested use would be different. If the product were tested as it would actually be used by any subject, no one would know how the product actually performed because its variability of use would be too great.

4. External Validity

External validity refers to the extent to which the results of a specific study can be generalized to the population at large, or to other environmental conditions. Unfortunately, no experimental designs have built-in controls for threats to external validity [14, 91]. External validity can be subdivided into these two components: population validity and ecological validity [91, 99]. Population validity deals with the generalizability of the results of a study to the general population, whereas ecological validity concerns the generalizability of results of the study to be appropriate to other settings.

The most obvious way of assuring the external validity of the study is to have the identical study conducted independently at different geographic locations or among different subject populations [99, 113]. If similar results are obtained and the same conclusions drawn by different investigators at each site or among each population, the external validity of the study can be considered, to a high degree of probability, satisfactory.

With these thoughts in mind, let us return to the surgical scrub evaluation and some of the mechanics necessary to perform a successful evaluation. Recall that we are interested in the immediate, persistent, and residual antimicrobial effects of the test and control products. A completely randomized test should be designed. That is, each subject in the study must be equally likely to be assigned the test or control products to be used in the study.

In order to compare the two products for their immediate, persistent, and residual antimicrobial effectiveness, we need some value against which to assess any postwash reduction in microorganism populations. That value is a baseline reading—the microbial population numbers that normally reside on the surface of the hands—determined by sampling prior to product testing.

In this study, the immediate antimicrobial effects will be measured just after the surgical scrub is conducted, and the persistent antimicrobial effects will be measured 6 hours after the scrub. The residual antimicrobial effects will be measured over the course of 5 consecutive days of product use. Once again, by taking the time to sketch out basic experimental design procedures, one can clearly and simply address the requirements necessary to achieve the study's objectives [14]. This is represented using the following experimental design, which is a pretest/posttest design (Table 5.1). Schemata of this type can be designed to accommodate any

TABLE 5.1
Experimental Design for Surgical Scrub Evaluation

Baseline Measurement	Independent Variable (treatments)	Dependent Variables (microbial counts over both hours and days)					
		Day 1		Day 2		Day 5	
R_A O_1	A	O2	Immediate 6-hour	O3	Immediate 6-hour	O4	Immediate 6-hour
R_B O_5	B	O6	Immediate 6-hour	O7	Immediate 6-hour	O8	Immediate 6-hour

R = Completely randomized design; each subject is equally likely to be selected into group A or B.

Independent variables:
A = Treatment with test product (group A)
B = Treatment with control product (group B)

Dependent variables:
O_i = Microbial counts after treatment with product (A or B). $O_i = O_2, O_3, O_4, O_6, O_7,$ or O_8. Note: O_1, O_5 are the starting values, or the baselines. They are connected to the dependent variable as Reductions = Baseline $_{(O1, O5)}$– Colony Counts $_{(O2 – O8)}$

number of test products and are crucial to accurate conceptualization of what will actually happen in the testing process [15].

It is also important to know the distribution of the generated data [15]. Recall that the dependent or response variable is the variable one must measure after all other variables have been controlled. It is more often known, simply, as the *variable*. Typically, in a topical antimicrobial evaluation, the response variable consists of the microbial colony counts. A problem presented in the analysis of microbial colony counts is that they are exponential, not linear. The vast majority of statistical models are linear models and cannot be used to evaluate nonlinear data [92]. Therefore, the microbial counts should be transformed to a linear scale (most often, a $\log_{10}$ scale) to be evaluated statistically.

Also, in statistical designs, it is necessary to establish the levels of both the alpha (α) and the beta (β) error, so that the appropriate number of subjects to be tested and the number of replicate measurements to be taken in each sampling period, relative to the desired confidence level, can be determined [14, 84]. Recall that α-error (Type I error) is committed by rejecting a null hypothesis that is actually true, and β-error (Type II error) is committed by accepting a null hypothesis that is false. In other words, α-error occurs when one states that there is a difference between products, when there actually is not, and β-error occurs when one concludes that there is no difference between products, when there actually is. The easiest way to control both α- and β-errors is to use enough subjects (replicates) such that the possibility of both α- and β-errors is reduced. Otherwise, merely adjusting the α-error to a very small level for significance will only serve to increase the probability of β-error [2, 90].

The question often arises in topical antimicrobial evaluations whether the study should be single- or double-blinded. In single-blind situations, the study participants do not know if they are receiving the test product(s) or the control product, but the investigator does. In a double-blind study, neither the investigator nor the subjects know who is receiving which product. In the "real world," pressure on an investigator to evaluate the sponsor's product in the best light can impose a tacit biasing of the study. The use of a double-blind study can eliminate this biasing problem. Once the study is complete and the data are evaluated, the true product identities may be revealed. To protect against experimental bias, the various processes of the study should be compartmentalized so that no one person has total control of all aspects of the study [14, 93]. This is especially important in studies where there is double-blinding, but the products to be evaluated are easy to distinguish. This would happen, for example, when the test product is a chlorhexidine gluconate product and the reference product is an iodophor. It is very easy to distinguish the two, no matter how they are labeled.

Commitment to creating an efficient, statistically based experimental design will serve to provide a cost-effective model that will generate valid results. For this reason, it will be well worth the time and effort to perfect a design *before* commencing work on the study [30].

5. Is the Study Biased?

Subjectively, we all have opinions about how information provided to individuals could bias the results of a study. Interestingly, in these discussions, nothing objective has ever been offered to support those conclusions.

There is one important question: Has anything been done to affect the objective study results?

Examine the baseline data to see what happened. There are four things that should be checked. They are so important, they are listed again.

1. Perform an EDA on the baseline data of all groups combined in the study. Specifically, do a stem-and-leaf and a letter-value display and see if the data look biased. Perform the analysis with all the subjects' data included in the study or with only the subjects who achieved the required $\log_{10}$ counts. If done this way, there will not be a normal distribution. For example, if you use only the people who passed the 1.5×10^5 $\log_{10}$ requirement, the data will look like Figure 5.3.
2. Is the standard deviation, s, the same as it has been in former studies? If it is larger—substantially larger—there is a problem.

Required 1.5×10^5 baseline

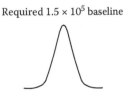

FIGURE 5.3   Distribution.

3. Were the study participants sufficiently randomized? Find out exactly if and how this was done. If one-quarter of the subjects were cheating, where would they have been placed—test product, control product, or both products equally?
4. Look at the control group of the study. Is there any noticeable difference from other previous controls for studies of this type? The mean and the standard deviation should be the same as in other studies of this type.

This four-step evaluation should be accomplished before the data are analyzed.

6. Hawthorne Effect

In experiments involving human subjects, a great many subtle influences can distort research results [94]. If individuals are aware that they are participating in an experiment, this knowledge may alter their activities, and thereby invalidate the experiment. Studies carried out at the Hawthorne, Illinois, plant of the Western Electric Company first called attention to some of these factors. In one of the studies, the illumination of three departments in which employees inspected small parts, assembled electrical relays, and wound coils was gradually increased. The production efficiency in all departments generally increased as the light intensity increased. Experimenters found that upon decreasing the light intensity in a later experiment, however, the efficiency of the group continued to increase slowly, but steadily. Further experimentation with periodic rest periods and variations in the length of working days and weeks were also accompanied by gradual increases in efficiency, whether the changes in working conditions were for the better or for the worse. Researchers concluded that the attention given the employees during the experimentation was the major factor leading to the production gains. This phenomenon has since been referred to by psychologists as the *Hawthorne effect*. The factory workers who carried out the same dull, repetitive tasks month after month were stimulated and motivated by the attention and concern for their well-being displayed by the research workers. A new element had been added to their uneventful existence—not the illumination or the other variables that the researchers were studying, but the researchers, themselves.

The term *Hawthorne effect* has come to refer to any situation in which the experimental conditions are such that the subject is aware of participating in an experiment, is aware of the hypothesis, or is receiving special attention which tends to improve performance. Certainly, many educational experiments report changes and improvements that are due primarily to the Hawthorne effect. For example, research in which one group of teachers continues with the same teaching methods it has previously employed, while another group is trained in a new method, will usually result in changes in teacher performance or student achievement favorable to the new method. Many school districts, in the process of trying out new methods, set up 1-year experiments in which a new method is introduced to a limited number of pupils. The results of such experiments are certainly influenced by the Hawthorne effect, because teachers usually approach a new method with renewed enthusiasm, and the students, aware that they are being taught by a new and different method, are likely to display more interest and motivation than usual. The influence of the

Hawthorne effect can be expected to decrease as the novelty of a new method wears off, so studies extending over a period of 2 or 3 years are considered more reliable for evaluating the effectiveness of a new technique.

The most common research strategies employed to assess the magnitude of the Hawthorne effect (and thereby neutralize its effect on the experiment) employ various kinds of control groups [95]. For example, suppose that a control group C is added to an experimental group E. Then the difference in outcome of the dependent variable between control group C and a nontreatment control group N would represent the Hawthorne effect, and the difference between control group C and experimental group E would represent treatment effect with the Hawthorne effect removed.

It is interesting that attempts to manipulate the Hawthorne effect experimentally often have failed to produce any effect. Nevertheless, there is much indirect evidence that it operates in many studies [14]. The researcher should take steps to reduce any special attention given the subjects, the novelty of experimental treatments, and the awareness of participation in a research project, because any of the circumstances may contribute to the effect. In any event, such precautions will improve the research design, whether the Hawthorne effect might influence the study's outcome or not.

7. John Henry Effect

The legend of John Henry tells of a black railroad worker who pitted his strength and skill at driving steel railroad spikes against a steam driver that was being tested experimentally as a possible replacement for the human steel drivers. The *John Henry effect* refers to a situation often found in education research in which a control group performs above its usual average when placed in competition with an experimental group that is using a new method or procedure that threatens to replace the control procedure. This phenomenon is probably quite common in educational studies in which a conventional teaching methodology is being compared with a new methodology [3]. Teachers in the control group feel threatened by the new methodology and make a strong effort to prove that their way of teaching is as good as the new method.

This effect was named and described by Robert Heinrich in 1970 while reviewing studies that compared television instruction with regular classroom teaching. He found that the classroom teachers in the control group often made a maximum effort to ensure that their students' performance matched that of students who viewed televised instruction [7].

The John Henry effect in human studies probably reflects in part the competitive desire to prove that artificial intelligence (AI) can do just as well as those people who are being trained. It is also probable that persons who know that they are members of a control group feel psychologically threatened by a situation in which they perceive they are *preordained* to come out second best.

Since Heinrich's work, several studies have been conducted in which the John Henry effect appears to have operated, because unusual effort in the control group was documented, and control subjects matched or exceeded the performance of experimental subjects. Gary Saretsky, who conducted extensive study of this phenomenon, concluded that the John Henry effect is likely to occur when an innovation is introduced in such a manner as to be perceived as threatening to one's job, status, salary, or traditional work patterns. Saretsky provided convincing evidence that the John Henry effect resulted in a marked increase in achievement in control group labs when those labs were compared with labs in which performance-contracting was employed [96]. He obtained data on performance of the control subjects for 2 years prior to the experimental year. These data showed that, during the experimental year, control group gains in laboratory skills, as measured by standardized tests, were much higher than in the 2 preceding years. Because performance-contracting is very threatening to managers, it seems obvious that the managers made a very strong effort during the year of the experiment.

Obviously, the John Henry effect could be easily confused with the Hawthorne effect [96]. The two have somewhat opposite effects on an experiment, however, because the Hawthorne effect reflects the impact of being part of an experiment upon the experimental group's performance, whereas the John Henry effect reflects the impact upon the control group in experiments where the experimental group is perceived as competing with or threatening to surpass the control group.

8. Pygmalion Effect

This effect takes its name from a controversial study by Robert Rosenthal and Lenore Jacobson, reported in *Pygmalion in the Classroom* [97]. These researchers demonstrated that teachers' *perceptions* of a student's intelligence in some cases appeared to bring about changes in that student's intelligence test scores. Thus, the term *Pygmalion effect* has come to refer to changes in a subject's behavior that are brought about by the experimenter's expectations. The effect has been replicated in some studies, but not in others. In any case, the possibility that this effect can occur should alert researchers to the importance of *not* conveying their expectations to the subjects. This would likely have a strong effect on the performance of laboratory technicians.

C. Microbiological Methodology

As previously discussed, the investigator must be familiar with the microbial ecology of the skin to conduct these types of clinical studies properly. The skin surfaces provide a unique habitat for microorganisms, and knowledge of histology, physical features, and nutrient factors of the skin is important.

The knowledge of its histological structure can be an aid in understanding both the physical and nutritional characteristics of the skin relative to distribution of microorganisms [14]. The various surface features of normal skin that must be taken into account relative to microbial populations expected at specific anatomical skin sites include eccrine sweat glands, sebaceous glands, apocrine glands, and hair [13]. Physical characteristics of skin influencing microbial growth include pH, temperature, moisture, and the oxygen–carbon dioxide tension of the surface and within pores. Factors taken into account from a microbial nutritional perspective include the host's age, sex, race, diet, nationality, and body morphology (e.g., obesity), inasmuch as these influence both surface features and physical characteristics of the skin. Finally, a knowledge of those microorganism types that normally inhabit and colonize the skin surfaces is valuable to the investigator [14]. Although species commonly encountered will vary according to the host variables mentioned, as will their relative numbers and distribution on the host, the normal flora usually includes the coryneform bacteria; the Micrococcaceae, including *Staphylococcus* species; *Streptococcus* species; various Gram-negative bacteria, including *Escherichia coli*; *Mycoplasma* species; dermatophytic molds and yeasts; and viral particles, often transiently. Within this context, the investigator can determine the best culture medium on which to grow the microorganisms that will be encountered at the anatomical sites sampled. Also, what dilution levels will be employed in testing and the appropriate incubation temperatures and periods must be established. Again, these kinds of questions must be addressed before beginning the study.

In the design of topical antimicrobial evaluations, the intended product use must be determined and in-use conditions simulated, insofar as possible [13, 16]. For example, if one wants information concerning the antimicrobial properties of a new surgical scrub product, inoculating the hands with an indicator microorganism such as *Serratia marcescens* may not be the most valid approach. Assessment of reductions in normal flora residing on the subject's own skin surface is probably better, since it more closely approximates the actual conditions.

D.   STATISTICAL METHODS USED TO EVALUATE THE STUDY

It is important that statistical models (linear regression, analysis of variance, analysis of covariance, Student's *t*-test, or other) be selected to comply with the experimental design [14]. If the investigator does not do this beforehand, it has the same effect as noncompliance with any protocol and may invalidate the entire study.

Although the exact statistical model to be used within the framework of the experimental design often depends on the distribution of the data generated (normal, skewed, bimodal, exponential, binomial, or other), the use of EDA can help the investigator not only select the appropriate statistical model, but develop an intuitive feel for the data before the actual statistical analysis occurs [98]. The investigator actually sees the shape of the data distribution in terms of graphic displays, can identify any trends that may not be obvious from the unprocessed data, and on that basis, can select an appropriate statistical model.

It is also important to identify the data-handling procedure in the study design. If prewritten software packages, such as Statistical Package for the Social Sciences (SPSSx), Biomedical Data Program (BMDP), Statistical Analysis System (SAS), or MiniTab™, are used, the software program must be configured. Each program requires some careful thought because the configuration must reflect, for example, how the samples will be taken, how data will be blocked, and what contrasts will be used [14]. In addition, the data input arrays are arranged to assure that the keyed input data file is compatible with the software's input capacity. Understandable graphic display formats will aid the programmer in presenting and then interpreting even the most complex data analyses.

E.   NEW PRODUCT DEVELOPMENT APPLICATIONS

Now, let us change our focus to the actual development of new products for the antimicrobial market. We will begin with a common problem in the industry—that of attempting to produce one antimicrobial product to meet multiple needs [13].

A standard practice of manufacturers in formulating topical antimicrobial products is to recommend their product as a surgical scrub, as a preoperative skin preparation formulation, and as a healthcare personnel handwash. The reason for this is simple. It takes less research and development work to offer one product covering all three categories than to develop a specific surgical scrub product, a specific preoperative skin preparation formulation, or a specific healthcare personnel handwash product.

This probably will change because of a very fundamental error in this reasoning. Each product category has its own unique requirements, and the topical antimicrobial product should be designed with these requirements in mind to meet its intended purpose [14, 30]. For example, surgical scrub products, in general, are too harsh and potentially irritating to the skin to serve well as a healthcare personnel handwash, which requires repeated use over the course of a day. The main function of a healthcare personnel handwash formulation is removal of the transient microorganisms. A surgical scrub product must remove not only transient microorganisms, but also resident microorganisms [30]. The surgical scrub formulation, then, is "stronger" than necessary to meet the requirements of a good healthcare personnel handwash formulation. Although surgeons may scrub their hands two or three times a day in the surgical setting, it is not uncommon for healthcare personnel to wash their hands 25 to 30 times, as they interact repeatedly with patients. Used 25 to 30 times daily, a formulation designed as a surgical scrub frequently produces significant skin irritation to the hands and, for this reason, will likely not be marketable [13]. Hence, this market is vulnerable to a manufacturer who develops a formulation specifically targeted to meet the demands of the product.

Using this example again, to be marketable, the healthcare personnel handwash product must be effective in reducing transient microorganisms, and it must be non-irritating after repeated and prolonged use. Healthcare personnel have demonstrated a strong preference for products that are mild and nonirritating to their skin, as well as effective [14]. Generally, mildness can be built into the healthcare personnel handwash in three ways:

1. Proportionally reduce the amount of active ingredients such as chlorhexidine gluconate, iodophors, or alcohol contained in the product. For example, instead of using the customary 4% level of chlorhexidine gluconate (CHG) found in many surgical scrub formulations, a 2% concentration or less of chlorhexidine gluconate is used. Much of the fast-acting properties of the antimicrobials are provided by alcohol.
2. Add skin conditioners or emollients to the formulation, thereby counteracting the irritating effects of the active compounds and making the product more gentle and mild to the skin.
3. Use a combination of these two methods; that is, reduce the active ingredient levels and add emollients and skin conditioners to the formulation.

Often, suppliers are not the manufacturers of the healthcare personnel products they distribute. As a result, they often feel that they are at a disadvantage in that they must accept what products are offered to them by the manufacturers. The suppliers have more options than often realized, however, even when using preexisting formulations. For example, they could collect samples from several manufacturers for use in a pilot study to determine the product most suitable for their needs. Through this approach, the optimum antimicrobial formulation—low in irritation potential, as well as antimicrobially effective—could be readily identified.

Suppose you want to evaluate both alcohol plus 2% CHG and alcohol plus 4% CHG formulations of four different manufacturers. An efficient statistical evaluation can be devised that will provide the information critical for determining the antimicrobial efficacy and the irritation qualities of the products. A potentially useful basis for evaluation of irritation potential is the repeated scoring of skin condition in terms of edema, dryness, erythema, and skin eruptions, perhaps employing a four-point rating system such as that shown in Table 5.1. A statistical model could then be designed based on these data to compare the irritation potential of the products.

A particularly useful nonparametric statistical model used for irritation evaluation is the χ^2 (chi square) statistic [99]. This model allows the detection of significant differences in the irritation produced by use of the products. The χ^2 statistic is statistically robust (reliable), yet accurate and precise in detecting true significant differences between products, as expressed by ordinal scale data produced from subjective evaluation of skin condition. Another possibility is using response surface methodology (RSM) to find the most suitable formulation. This procedure is extremely useful and powerful.

Let us now turn our attention from the specific example of evaluating a healthcare personnel handwash to the more general statistical strategy of evaluating new product formulations for which no known antimicrobial performance characteristics are available to the investigator.

1. Marketing Concerns

Although this is not often considered in the research and development effort, it is important. The most effective topical antimicrobial product is not of much value if there is no market. So the question becomes, "What are the market needs [14]?"

The topical anti-infective market, which mainly comprises healthcare personnel handwashes, surgical hand scrubs, and preoperative skin preparation formulations, is experiencing "new" product developments on a continual basis. An overview of the current advertising literature demonstrates this. But after reviewing these, are the developments really focused on "new" products or are they the standard products delivered in a new manner? The answer is "yes" to both these questions. There are new products being developed for the anti-infective market, as well as new systems of delivery and new users for standard products.

Let us now take a closer look at the research and development activity in the surgical scrub and preoperative skin preparation markets. As we have previously discussed the healthcare personnel handwash formulations in some detail, it will be omitted from this section.

a. Surgical Scrubs

Recall that surgical scrub formulations are designed to remove both the transient microorganism population, as well as a large proportion of the normal endogenous microorganism population residing on the hands. To be considered effective, surgical scrub formulations must demonstrate both immediate and persistent antimicrobial effectiveness (up to 6 hours postscrub) and, optimally, demonstrate a *residual effect* by becoming more effective antimicrobials with repeated use over time because the active ingredient(s) are absorbed directly into the skin [2].

Over the years, the povidone-iodine market share has been significantly eroded by chlorhexidine gluconate formulations because of the latter's residual properties, and recently there has been interest in developing low-level (2% or 3%) chlorhexidine gluconate formulations for the surgical scrub segment. Presently, there is no longer a scrub procedure provided for these products because they do not need any. They are, instead, made with alcohol for fast kills and have a backup drug to take over after the alcohol dries.

b.   Preoperative Preparative Solutions

The preoperative skin preparative solution is designed to both degerm an intended anatomical surgical site and provide a high level of persistent antimicrobial activity (up to 6 hours postpreparation) [14]. Many products have been developed that have antimicrobial properties lasting much longer than 6 hours; however, because the FDA has not addressed this, there are no requirements for evaluations beyond 6 hours. In the past, preoperative skin preps have been the domain of the iodine products, but these are being aggressively challenged by several chlorhexidine gluconate formulations. In addition, formulations providing long-term persistent antimicrobial activity (up to 96 hours after preparation) are likely to be introduced.

Generally, this changed when alcohol was added. The alcohol allows for fast dry times and high, immediate kill rates. There is usually another drug, such as CHG, that provides the persistent and residual antimicrobial effects.

2.   Other Considerations

Although there is some market activity in developing and introducing new active ingredients for the healthcare personnel handwash, surgical scrub, and preoperative skin preparative products, most of the efforts will be focused on using the common antimicrobials (e.g., iodophors and chlorhexidine gluconates) as the active ingredients, but applied in new and novel delivery systems [2].

Increasingly, healthcare personnel formulations will be purposefully developed as healthcare personnel products, not relabeled surgical scrubs. There are several products in development that can get high/low reductions on later application. Surgical scrub formulations introduced in the 1990s will, again, tend to be modifications of existing products. The preoperative skin preparative solution area will experience the most activity, with new product delivery systems.

A *full-body shower wash* is used off-label in conjunction with preoperative preparation regimens [56]. The full-body shower wash will be employed to reduce the microbial baseline counts on the patient's body before being prepared for surgery. The preoperative skin preparation formulations will then have a reduced microorganism population to contend with, making it more effective in its intended use. It is unfortunate that the FDA has not looked into this area.

We are often asked, "How does one market topical anti-infective products?" But there is no "trick" to successful marketing of topical anti-infective products; only thorough planning and implementation of a creative marketing program are required [100]. An analysis of new product development programs generally would show that the programs themselves are well thought-out and ultimately very successful. Frequently, however, these programs are based on good intentions, but do not meet the actual market requirements.

Not uncommonly, companies will have problems when they enter the topical antimicrobial market with a product that is essentially the same as a competitor's product, and in doing so, face an uphill battle trying to market a product that is not unique [20, 100]. There are many suppliers of surgical scrub, healthcare personnel, and preoperative skin preparative formulations, and most of their products are basically identical to one another. For example, the vast number of alcohol and CHG products used for surgical scrub formulations are essentially identical 4% CHG solutions. Since there is no real difference between these products, the major selling point ultimately becomes the sales price.

This situation is particularly unfortunate when the industry has such tremendous potential for new, innovative products, as well as new applications for existing ones. For example, there are a number of alcohol-based products that are rinseless. There is also a need for procedures to be used in conjunction with the preoperative skin preparative process to enhance the total antimicrobial effect. If a full-body shower wash product application significantly reduces the microbial flora residing on the skin, the follow-up preoperative skin preparation procedure will likely be more efficacious than it would be alone, since fewer microorganisms are left with which to contend.

It is important to appreciate that when new product evaluation methods are designed for new products and receive the FDA stamp of approval, these methods will very likely be adapted industrywide as the procedures for evaluation of future me-too products.

II. CONCLUSION

We have concentrated on a broad, integrative perspective to evaluating topical antimicrobial products in this chapter. We have discussed these evaluations from experimental design, microbiological, biostatistical, and marketplace perspectives. Based upon experience, it seems that committing to such a holistic, detailed approach to program design significantly aids in evaluating products, and in moving successfully through the FDA approval process because of the straightforward, concise, and unambiguous manner of conducting valid topical antimicrobial clinical trials.

6 Current Topical Antimicrobials

In this chapter, we discuss various aspects of the topical antimicrobial products currently in common use in medical, food service, and consumer (personal hygiene) markets. The antimicrobial products of primary interest include iodine complexes (aqueous iodophors and tinctures), aqueous formulations of chlorhexidine gluconate (CHG), triclosan, parachlorometaxylenol (PCMX), alcohol formulations, tinctures of CHG, and quaternary ammonium products [2, 31]. Commonly, alcohol is formulated purely or blended with other drugs, such as iodine and CHG, to increase its performance benefits.

I. IODINE COMPLEXES

Iodine in its pure form is relatively insoluble in water without a solubilizing agent, but it dissolves well in various alcohols to provide an iodine tincture [64]. Tinctures of iodine are used primarily as antiseptics.

By far, the most common form of iodine for use as a topical antimicrobial is the iodophor [101]. Iodophors are complexes of elemental iodine (tri-iodine) linked to a carrier having several functions: (1) increased degrees of solubility in aqueous solution for the iodine, (2) provision of a sustained release reservoir of the iodine, and (3) reduced equilibrium concentrations of free iodine. The most commonly used carriers are neutral polymers, polyacrylic acids, polyether glycols, polyamides, polysaccharides, and polyalkalines.

The most commonly used iodophor is povidone iodine, a compound of 1-vinyl-2-pyrrolidinone polymer with available iodine ranging between 9% and 12% (*United States Pharmacopeia XXIII*) [102]. The chemistry of aqueous solutions of iodophors is complex. Based on our current understanding, various electronic and steric effects are responsible for the interaction observed between polymeric organic carrier molecules and iodine (Equations 6.1 to 6.5) [103].

$$I_2 + H_2°OI^+ + I^- \; (hydrolyticionization) \tag{6.1}$$

$$H_2 + OI^+ °HOI + H^+ (dissociation\ of\ H^+) \tag{6.2}$$

$$HOI°OI^- + H^+ (dissociation\ of\ HOI) \tag{6.3}$$

$$3HOI°IO_3^- - 2I^K 3H^+ \tag{6.4}$$

$$I_2 + I - {}^\circ I_3 - (formation\ of\ triiodide) \tag{6.5}$$

This interaction permits high degrees of variation in the equilibrium concentrations of iodine. Hence, it is assumed that in interaction with low molecular oxygen compounds such as amides, esters, ketones, or ethers, donor–acceptor complexes are formed with the iodine playing the part of the acceptor.

$$\zeta = O + I_2 {}^\circ \zeta = O - I - I \tag{6.6}$$

The relatively low concentration of free molecular iodine in povidone iodine concentrates is a reason that it may not be capable of autosterilization [103]. For example, *Pseudomonas* spp. and *Burkholderia cepacia* have reportedly been isolated from iodophor concentrates. Apparently, the bacteria are protected by biofilms [104].

The amount of free molecular iodine is highly relevant to the antimicrobial effectiveness of iodophors. The differing composition of pharmaceutical additives (e.g., detergents and emollients) that have iodine-complexing properties, as well as the ratio of free iodine to total iodine in various proprietary formulations, can result in great differences in the concentrations of free molecular iodine available for binding. Hence, the antimicrobial index of importance for a product is the *total available iodine* [105].

Iodine preparations play a variety of roles as skin disinfectants [83]. For example, iodophors are commonly used as preoperative skin preparations, as surgical scrub solutions, and as preinjection preparations, as well as preinsertion preparations for venous and arterial catheters. Iodophors have also been used successfully as therapeutic agents in treating wound infections, including those encountered in burn patients, and for disinfection of medical equipment, such as sutures, catheters, scalpels, plastics, rubber goods, brushes, and thermometers. Finally, iodophors have been used successfully as an antimicrobial treatment for drinking water, swimming pool water, and waste water.

A.   RANGE OF ACTION

Iodophors and tinctures of iodine provide excellent immediate antimicrobial action against a broad range of viruses, both Gram-positive and Gram-negative bacteria, fungi, and various protozoa [103]. In fact, almost all important human disease microorganisms, including enteric bacteria, enteric viruses, protozoan trophozoites and cysts, mycobacteria, spores of *Bacillus* spp., and *Clostridium* spp. and many fungal species, are susceptible to free iodine. It should be noted, however, that exposure times and concentrations of available iodine required vary (Table 6.1) [103].

In topical application to skin surfaces (e.g., hands and body surfaces in the inguinal, abdominal, anterior cubital, subclavian, and femoral regions), iodophors and tinctures of iodine providing at least 1% available iodine demonstrate effective immediate and persistent antimicrobial properties [103]. It should be noted in general, however, that neither provides residual antimicrobial action [14, 56].

TABLE 6.1
Some Recommended Applications for Iodine-Based Antimicrobials

Scope of Application	Concentration	Conditions	Exposure Time	Disinfective Result	References
General germicidal action	1:20,000	Absence of organic matter	1 min	Most bacteria are killed	[93]
	1:20,000	Absence of organic matter	15 min	Wet spores are killed	[93]
	1:200,000	Absence of organic matter	15 min	Will destroy all vegetative forms of bacteria	[93]
Disinfection of skin	1% tincture	—	90 sec	Will kill 90% of the bacteria	[93]
	5% tincture	—	60 sec	Will kill 90% of the bacteria	[93]
	7% tincture	—	15 sec	Will kill 90% of the bacteria	[93]
	1% aqueous I_2-solution	Skin of hands	20 min	Inactivation of rhinovirus	[92]
	2% aqueous I_2-solution	Skin of hands	3 min	Inactivation of rhinovirus	[92]

II. CHLORHEXIDINE GLUCONATE

CHG was first synthesized in 1950 by ICI Pharmaceutical in England [107]. CHG was found to have high levels of antimicrobial activity, but relatively low levels of toxicity to mammalian cells [107, 108]. Additionally, CHG has a strong affinity for skin and mucous membranes. As a result, CHG has been used as a topical antimicrobial for wounds, skin prepping, and mucous membranes (especially in dentistry), where it provides, by virtue of its proclivity for binding to the tissues, extended antimicrobial properties. CHG also has value as a product preservative, including ophthalmic solutions, and as a disinfectant of medical instruments and hard surfaces.

CHG is a cationic molecule that is generally compatible with other cationic molecules such as the quaternary ammonium compounds [107]. Some nonionic substances such as detergents, although not directly incompatible with CHG, may inactivate the antimicrobial properties of CHG, depending upon the compound and concentration levels. CHG is *incompatible* with inorganic anions, except in very dilute concentrations, and may also be incompatible with organic anions present in soaps containing sodium lauryl sulfate, and with a number of pharmaceutical dyes [101, 105, 107].

The antimicrobial activity of CHG is pH-dependent, with an optimal use range of 5.5 to 7.0, a nice match with the body's usual range of pH. The relationship between antimicrobial effectiveness of CHG and pH varies with the microorganism, however [107]. For example, its antimicrobial activity against *Staphylococcus aureus* and *Escherichia coli* increases with an increase in pH, but the reverse is true for *Pseudomonas aeruginosa*.

The antimicrobial activity of CHG against vegetative forms of both Gram-positive and Gram-negative bacteria is pronounced [107, 109]. It is generally inactive relative to bacterial spores, except when they are exposed at elevated temperatures. Mycobacteria—acid-fast microorganisms—are reportedly inhibited, but not killed, by CHG in aqueous solutions. A variety of lipophilic viruses (e.g., herpes virus, HIV, influenza virus) are rapidly inactivated by exposure to CHG. Finally, certain fungi, particularly those in the yeast phase, are sensitive to CHG.

A. ANTIMICROBIAL ACTION

At relatively low concentrations, CHG exerts bacteriostatic effects on bacteria. At higher concentrations, CHG demonstrates rapid bactericidal effects. However, the precise effects vary from species to species and as a function of concentration of CHG [107, 109].

The microbicidal effects occur in a series of steps related to both cytological and physiological changes, which culminate in the death of the cell [107, 108]. CHG is known to have an affinity for bacterial cell walls and is absorbed into certain phosphate-containing cell wall compounds. By this process, CHG is thought to penetrate the bacterial cell wall, even in the presence of cell wall molecular exclusion mechanisms. Once cell wall penetration has occurred, the CHG is attracted to the cytoplasmic membrane. Upon penetration of the cytoplasmic membrane, low-molecular-weight cellular components (e.g., potassium ions) leak out of the membrane, and membrane-bound enzymes such as adenosyl triphosphatase are inhibited. Finally, the cell's cytoplasm precipitates, forming complexes with phosphated compounds, including adenosine triphosphatase (ATP) and nucleic acids. As a rule of thumb, bacterial cells carry a total negative surface charge [107, 108]. It has been observed that, at sufficient CHG concentrations, the total surface charge rapidly becomes neutral and then positive. The degree of the shift in the charge is directly related to the concentration of CHG, but reaches a steady-state equilibrium within about 5 minutes of exposure. The rapid electrostatic attraction between the cationic CHG molecules and the negatively charged bacterial cell surface contributes to the rapid reaction rate, that is, the rapid bactericidal effects exerted by CHG.

CHG at antiseptic concentrations (0.5 to 4%) demonstrates a high degree of antimicrobial effect, both -static and -cidal, on vegetative phases of Gram-positive and Gram-negative bacteria, but it has little sporicidal activity [107]. While there have been concerns that prolonged use of CHG could lead to reduced sensitivity and, ultimately, the development of resistant strains of bacteria, this concern has not been verified, even upon prolonged and extensive use [107, 108]. There is no evidence that plasmid-mediated antimicrobial resistance, particularly common in Gram-negative bacteria, has developed. This has been borne out in studies of common indicator species such as *Escherichia coli, Pseudomonas aeruginosa, Serratia marcescens,* and *Proteus mirabilis*. Although several researchers have reported a reduced sensitivity to CHG among certain methicillin-resistant strains (MRS) of *Staphylococcus aureus*, at clinical-use concentrations, this concern has not been substantiated. MRS and non-MRS strains are equally susceptible to CHG.

Because viruses have no synthetic properties of their own, the action of CHG is restricted to the nucleic acid core or the viral outer coat. Some viral coats consist of protein, and others, lipoprotein or glycoprotein. Outer envelopes that enclose the vital coat of some viruses are mainly of lipoprotein.

In general, CHG is a highly effective antimicrobial in its immediate, persistent, and residual properties [107, 108]. Higher concentrations of CHG (up to 4%) provide excellent immediate and persistent action, with the added benefit (when used repeatedly over time) of good residual effects. CHG at lower concentrations (less than 0.5%) provides antimicrobial action comparable to PCMX and triclosan.

CHG, by virtue of its residual antimicrobial properties, may be useful for full or partial bodywashes prior to elective surgery. If the product is used repeatedly over the course of at least 3 to 5 days, the resident microbial populations are reduced by about 3 logs [56]. Hence, when a person undergoes surgery, the remaining microbial populations residing on the proposed surgical site have been very significantly reduced, leaving far fewer microorganisms to be eliminated by preoperative prepping procedures.

III. TINCTURES OF CHG AND ALCOHOL

Currently, there is much interest in alcohol tinctures of CHG [2, 31]. These alcohol–CHG products may prove to be highly effective for use as preoperative skin preps, surgical scrubs, and healthcare personnel handwash formulations. Additionally, tincture of CHG may be useful as both a preinjection and prearterial/venous catheterization prep. Preparations of alcohol–CHG combine the excellent immediate antimicrobial properties of alcohol with the persistence properties of CHG to provide a clinical performance superior to either alcohol or CHG alone. The current response to CHG and alcohol is that CHG is slow-acting; however, it does have persistent properties, while alcohol is fast-acting but has no persistent properties (Figure 6.1).

Tinctures of CHG are used when one wants a fast reduction in microorganisms and then to be held down for a considerable time. Applications include peripherally inserted central catheter (PICC) lines and other such lines (IVs, midlines, central venous lines, and others).

IV. PARACHLOROMETAXYLENOL

PCMX is one of the oldest antimicrobial compounds in use, dating back to 1913. It has not been widely used as a surgical or presurgical skin preparation because of its relatively low antimicrobial efficacy [110].

Because of the absence of substantive data, the Food and Drug Administration (FDA) in 1972 did not designate PCMX as a safe and effective antimicrobial, and it was not formulated at the time for medical purposes such as surgical scrubs, preoperative skin preps, or healthcare personnel handwashes. The initial evaluations from

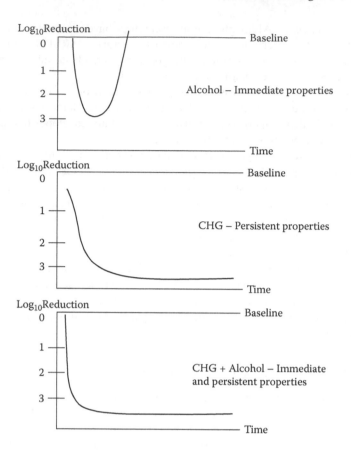

FIGURE 6.1   Properties of CHG and alcohol.

the FDA listed PCMX as a Category III product, meaning that there were not enough data to recognize it as both safe and effective as a topical antimicrobial [110].

Studies since 1980 have demonstrated PCMX to be safe for human use. After this determination, a number of companies became interested in developing PCMX for use as a topical antimicrobial, and later studies demonstrated PCMX to be an effective antimicrobial compound.

Current over-the-counter formulations demonstrate varying degrees of antimicrobial efficacy, depending upon the formulation. Generally, PCMX products achieve fair to good immediate effects and fair persistent effects, but like iodophors, they provide no residual antimicrobial activity. PCMX products currently are formulated mainly for healthcare personnel handwash application. These are effective in removing transient (contaminant) microorganisms from the hands and have low skin-irritation properties.

V.   ALCOHOLS

There is much debate concerning the effectiveness of alcohol as a skin antiseptic [110, 111]. The antimicrobial efficacy of alcohol is highly dependent upon the concentration used, as well as the moisture level of the microbial environment treated. The short-chain, monovalent alcohols—ethanol and isopropanol—are probably the most effective for skin disinfection because they are highly miscible with water, have low toxicity and allergenic potential, are fast-acting, and are microbicidal as opposed to microbiostatic.

The microbicidal activity of the alcohols is largely a function of their ability to coagulate proteins. The literature suggests, however, that microbicidal effects of alcohols are also a result of their solubility in lipids. Protein coagulation takes place on the cell wall and the cytoplasmic membrane, as well as among the various plasma proteins.

Alcohols are generally inactive against bacterial spores. And although there is much controversy in the literature concerning the efficacy of alcohols against viruses, there appears to be general agreement that enveloped, lipophilic viruses are more susceptible to inactivation by alcohols than are "naked" viruses. Last, the fungicidal properties of alcohols vary among fungal species, but in general, alcohols demonstrate a relatively high degree of mycidal/-static activity.

Although alcohols, as topical skin disinfectants, provide excellent immediate antimicrobial activity, they have virtually no persistent or residual properties [110]. Hence, their value as surgical scrubs and preoperative skin preps is seriously limited [2, 31]. Alcohols have been shown to provide adequate results as healthcare personnel handwashes or preinjection skin preps in removing or killing transient microorganisms, but when used at strengths of 70% and greater, they tend to be drying to the skin, resulting in significant irritation.

VI. TRICLOSAN

Triclosan, like PCMX, provides varying degrees of antimicrobial efficacy, depending upon the specific formulation and the species of organism tested [2, 31]. Triclosan has been formulated for a wide range of applications and is currently used in healthcare personnel handwash formulations, in the food industry to cleanse workers' hands, and in consumer product lines, including hand soaps, shower gels, and body cleansers. Triclosan, like PCMX, provides fair immediate and persistent effects, but no residual action.

VII. QUATERNARY AMMONIUM COMPOUNDS

Quaternary ammonium chlorides (QACs) occupy a unique niche in the world of antimicrobial compounds. Rather than being a single, well-defined substance, as is the case for many antimicrobially active ingredients, QACs are composed of a diverse, eclectic collection of substances that share a common molecular structure containing a positively charged nitrogen atom covalently bonded to four carbon atoms. This carbon–nitrogen structure is responsible for the name of these antimicrobial compounds and also plays a dominant role in determining their chemical behavior.

The first reports of quaternary ammonium compounds with biocidal activity appeared in 1916. Since that time, QACs have grown in popularity and been utilized extensively as active ingredients in many types of products, including household cleaners, institutional disinfectants, skin and hair care formulations, sanitizers, sterilizing solutions for medical instruments, preservatives in eye drops and nasal sprays, mouthwashes, and even in paper processing and wood preservatives. As a group, QACs are effective across a broad spectrum of microorganisms, including bacteria, certain molds and yeasts, and viruses. However, the specific activity of QACs is as diverse as their range of chemical structures. QAC antimicrobial effectiveness is highly formulation dependent, because a variety of compounds may affect QAC activity. Some components reduce the QAC efficiency, while others may synergize their activity or expand the spectrum of affected microorganisms. This fact has led to some confusion and apparent contradiction in the published literature as to the actual effectiveness of QACs in their role as antimicrobials.

In addition to their antimicrobial activity, QACs also behave as surfactants, assisting with foam development and cleansing action. They are also attracted to the skin and hair, where small amounts remain bound after rinsing. This contributes to a soft, powdery feel to the skin, unique hair-conditioning effects, and long-lasting, persistent activity against microorganisms. These various attributes and multifunctional roles of QACs appeal to formulators and are responsible for their incorporation into many consumer products.

7 General Statistics

After performing a study and everything was done properly, then what does the study show? In other words, what are the results? This section covers few things that only a statistical analysis can reveal.

The first thing that is important is to keep the analysis the same as has been done in the past, so there is consistency. The second thing to do is follow the protocol specifically as to how the data will be analyzed. This section describes how to perform a statistical analysis.

I. SAMPLE SIZE DETERMINATION

In the Food and Drug Administration's (FDA's) TFM (on page 31446, Section F, [112]), there is a formula for the number of subjects to use per arm for a study. It is

$$n \geq \frac{2s^2\left(z_{\alpha/2} + z_\beta\right)^2}{d^2},$$

where:

n = Sample size per arm of the study; that is, how many subjects will be used per product.

s^2 = Variance of the population. Use the most extreme of the two variances determined through the pilot study.

$z_{\alpha/2}$ = This represents the alpha (α), or Type I error. It is the probability of stating that a false H_0 hypothesis is true. It uses the z tables, meaning it is for an infinite population size. That is, it is not a t-value, which takes into account the population size. The alpha value is $\alpha = 0.05$, or $\alpha = \dfrac{0.05}{2} = 0.025$ for two-tail tests. This provides a $z_{\alpha/2}$ value of 1.96.

z_β = This is the β (beta) value for Type II error. It is the probability of stating that a false H_1 is true. It is set at $1 - \beta =$ Power of the statistic. If Beta = 0.20 $(1 - 0.20 = 0.80)$. This has a z value of 0.842.

d = The difference one wants to be able to determine. It is assumed that $d = 0.5 \log_{10}$.

The formula can be modified for more than two comparisons:

$$n \geq \frac{x s^2\left(z_{\alpha/2} + z_\beta\right)^2}{d^2},$$

where x = number of products tested. All the rest of the values have the same meaning.

A pilot study should be done to get a representation of the data to see if the product will pass the TFM requirements, and also to find the variances, or s_i^2 values. Remember, use the largest of the s_i^2 values in this formula as the variance.

Example

A pilot study was performed using four products:

1. Test product (alcohol/ chlorhexidine gluconate [CHG])
2. Control alcohol
3. Control CHG
4. Reference product (alcohol/CHG)

For this pilot study, the four products produced these variances (s^2):

	s^2
1	0.59
2	0.79
3	0.81
4	0.92

For the variance of the sample size calculation, the most extreme value is used: $0.92 = s^2$. This helps protect against using too few samples, which would be disastrous, for the final study would be too small and the study would be invalid.

The number of samples would be

$$n \geq \frac{4(0.92)(1.96 + 0.842)^2}{5^2}$$

$$n \geq 115.5697$$

$$n \geq 116 \text{ per arm of the study.}$$

In other words, for a surgical scrub and healthcare personnel handwash, the study would require 116 subjects × 4 products = 464 subjects total. For a preoperative skin preparation study, the study would require (116 subjects × 4 products) ÷ 2 sides = 232 subjects. Remember, a preoperative skin preparation uses the left and right sides of a subject, so each subject receives two products. For a preoperative skin preparation study, the randomization would be different.

Suppose three products are used. How many subjects would be used to get at least 93 subjects per arm? First, determine how these products should be configured. Arrange the three products

A   B   C

into

Subject (Left and Right Sides)

to determine the configuration:

Block	Subject	Configuration
1	1	A B
1	2	A C
1	3	B C

We find that one block of three people randomized the whole process in that each person gets two products. Three is the magic number, and the number of people per product will have to be in terms of the block size.

$$3$$
$$6$$
$$9$$
$$\vdots$$
$$36$$

The values that will not work are not evenly divisible by 3. For example,

$$5$$
$$11$$
$$19$$
$$\vdots$$
$$37$$

Next, from a pilot study recently performed, we choose the three products with the highest variances:

Product	s^2
A	0.82
B	0.98
C	0.84

The highest s^2 value was Product B, with a value of 0.98. Then we calculate the sample size:

$$n \geq \frac{3(0.98)(1.96 + 0.842)^2}{0.5^2}$$

$$n \geq 92.33$$

$$n \geq 93 \text{ subjects per product.}$$

We need 93 subjects per product, and there are three products and two sides. The formula for this is as follows (note that the total number of people required must be an even value):

$$\frac{(\text{minimum number of people required per product})(\text{number of products tested})}{2 \text{ sides}}$$

= total number of people required for a preoperative skin preparation study. (The number of people required must be an even number.)

Let us work an example, taking an uneven number to illustrate. Using 93 people and 3 products, we have:

$$\frac{93 \times 3}{2} = 139.5 \text{ subjects (A fraction will not work; this must be an even number.)}$$

Try 94 subjects.

$$\frac{94 \times 3}{2} = 141 \text{ subjects, 94 subjects per product (This is an even number, so it will work.)}$$

Perform the study based on 141 subjects, with 94 subjects per product.

What are the chances of the product actually passing the testing requirements? Let us look at two of the three types of tests:

1. Surgical scrub
2. Healthcare personnel handwash

For the surgical scrub (first application), after 1 minute these products must meet the 1.00 $\log_{10}$ reduction requirement with the lower confidence interval being greater than the $\log_{10}$ values (see Figure 7.1).

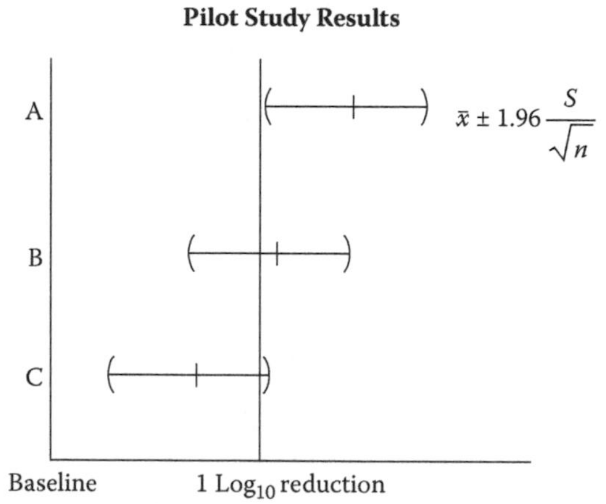

Pilot Study Results

FIGURE 7.1   Surgical handwash confidence intervals (day 1, immediate results).

Let us look at only one test day for the immediate results. For day 1, the result was 1.0 $\log_{10}$ reductions. This will keep our example simple. The wash times for the three products, A, B, and C, were 3, 2, and 1 minutes. Product A is the one that passed the test.

Had product A not passed, as long as the mean is more than the required $\log_{10}$ reduction, the product can be forced to pass by tightening the confidence intervals by adding more subjects. For example, product B had a mean of 1.2 $\log_{10}$, and the standard deviation (s^2) was 1.2, so increasing the number of subjects (n) will allow the product to pass by shrinking the confidence interval. We can see that its mean was greater than 1.0 $\log_{10}$, so it will pass, given that enough subjects are used.

$$\bar{x} \pm 1.96 \frac{s}{\sqrt{n}} \text{ *}$$

$$1.2 \pm 1.96 \frac{1.2}{\sqrt{116}}$$

$$0.98 \leq \mu \leq 1.42$$

The lower bound of 1.0 is intersected, so the test fails. We increase the sample size. We do this by trial and error (only the lower limit of the confidence interval is shown).

$$n = 130$$

$$x = 1.2 - 1.96 \frac{1.2}{\sqrt{130}}$$

---

* Note that the Food and Drug Administration (FDA) uses the standard deviation as the value here. If there were >2 product evaluations, the mean square error (MS_E) calculated by analysis of variance (ANOVA) would be the s^2 or MS_E, one or two values where $\log_{10}$ reductions are taken.

Theory 1: There is a subtotal of $\bar{x}_{BL} - x_{Time}$ = reductions. This is composed of two values. The formula should be $\bar{x}_{BL} - \bar{x}_{Time} \pm 1.96 \sqrt{\frac{2MS_E}{n}}$. Because the two values are taken, the variance is 2 MS_E.

The problem with this is that the MS_E of the baseline is narrow. When the time is taken at 1 or at application 3, the variance expands.

(----------) = baseline
(----------) + (----------) = baseline + application 1
(----------) + (----------) + (----------) = baseline + application 1 + application 2

There is no way to take the 5th measurement without taking the 4th, 3rd, 2nd, and 1st measurements. This is a problem. If one compares the variances equally, one finds that they are not equal.

Theory 2: Do the $\bar{x}_{BL} - \bar{x}_{Time}$, and count it as one measurement. That is, $\bar{x}_{BL} - \bar{x}_{Time} = \bar{x}_{\text{Log Reduction}}$. $\bar{x}_{\text{Log Reduction}} \pm 1.96 \sqrt{\frac{MS_E}{n}}$. This is done by converting the primary way. Because of vast knowledge of previous studies, we know that this conversion works.

$$x = 0.9937 \not> 1 \ \log_{10}, \text{ so it will not pass.}$$

$$n = 160$$

$$x = 1.2 - 1.96 \frac{1.2}{\sqrt{160}}$$

$x = 1.01 < 1.0 \ \log_{10}$, so it passes, but barely. You can choose to do this, or you could increase the wash time to 2½ minutes. It is important, however, that this is done so that no other companies sell their products as quicker to use, gaining a competitive advantage. Then perform another pilot study to check the $\log_{10}$ reductions.

For product C, if a pilot study that results in a confidence interval of less than the FDA TFM requirement of 1.0 $\log_{10}$, and the mean value is also less than the FDA value, do not perform the study. It will always be less than the required value. In this case, you must go back to the drawing board and use more milliliters of product, use a greater percentage of product (1% to 2%), use more friction, and/or a longer application time.

II.   WHAT TYPE OF ANALYSIS SHOULD I USE TO ENSURE THE STUDY HAS BEEN DONE CORRECTLY?

Before beginning the analysis, it is wise to check the baseline data for normality. For example, how do you know the subjects have been randomly sampled? That is, do the data fit a normal distribution? Are the data skewed left or right, and are there any outliers or extreme values that do not fit the distribution?

There are four aspects of the baseline data. I suggest checking these, regardless of which study is being analyzed.

A.   EXPLORATORY DATA ANALYSIS (EDA)

Perform EDA on the baseline data of all subjects in the study. Specifically, prepare a stem-leaf display and a letter-value display. Make the transformation to $\log_{10}$ scale (otherwise, the data will be skewed). For example, a 5.0 $\log_{10}$ minimum requirement of bacterial populations would look like Figure 7.2(a). This is okay, but what if the data looked like Figures 7.2(b) or 7.2(c)?

Sometimes, when only a small number of subjects are used, you will get results like (b) and (c). When you see multiple peaks, this would be a potential problem if the study were powered correctly. If the study were large enough, then it could represent the difference between males and females, or it could be differences in technicians' sampling and calculating, or differences in the media or testing days. This is why randomization is so important.

Suppose we are performing a healthcare personnel handwash. Figure 7.3 is a 95% Bonferroni confidence interval for the variances (s^2), graphed as standard deviations (s). None of these standard deviations are different from the others. Notice the baseline standard deviations are smaller than the wash applications. The wash standard

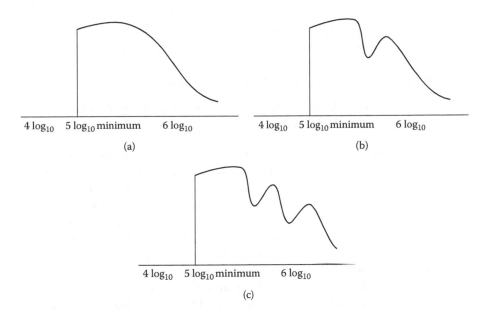

FIGURE 7.2 Distributions. a) abnormal distribution with 5 $\log_{10}$ minimum. b) bimodal distribution with 5 $\log_{10}$ minimum. c) trimodal distribution with 5 $\log_{10}$ minimum.

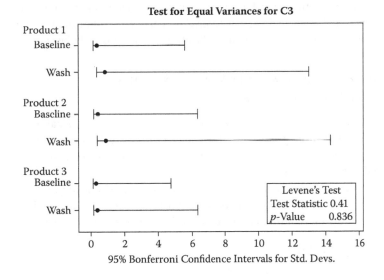

FIGURE 7.3 Test for equal variances.

deviations also contain the baseline standard deviations. That is, the wash had the baseline and wash standard deviations included; the wash n has a combination of these (Figure 7.4).

We see that the baseline is smaller for all the products than for the wash. This is because we have a baseline reading and Wash 1, which is the baseline plus Wash 1.

Baseline +	Wash 1	+ Wash 2	+ Wash n
(--------)			
(--------)	(----------)	(---------------)	(--------------------)

FIGURE 7.4   Standard deviation expands during the washes.

The more washes in the study, the greater the deviation. They are not categorized like this, however; they look like Figure 7.3.

1. *What value is an acceptable standard deviation?* It should be 1.0 $\log_{10}$ or less for healthcare personnel handwashes, surgical scrubs, and preoperative skin preparations. If it is greater than 1.0, there will be a problem in determining which product is better than the others. More subjects should be used or a serious review of the data and laboratory is recommended.
2. *Randomization.* Were the subjects randomized according to the study plan? If one quarter of the subjects were cheating, where would they be placed—test product, control product, or both?

    If the study were randomized, equal subjects would be placed in both groups, expanding the standard deviations, but not affecting which product did better.
3. *Control of the study.* To see if a study was run correctly, look at the control group. This group will generally have the same standard deviations across all studies. Does the control group have the same mean ($\bar{x}$) and standard deviation (s) that it has had in past studies? It is extremely important to check this because if this were acceptable, the study would probably be satisfactory.

B. COVARIANCE OR ANALYSIS OF VARIANCE STATISTICS: WHICH IS THE CORRECT WAY?

If you have three groups in a study with different baselines, this can be a problem. How can you tell if the products performed the same or differently, compared to the others? The obvious course of action is to perform an analysis of covariance to adjust the baseline counts. Or is it?

One of the requirements of analysis of covariance is that the regression slopes are the same for all the test products (see Figure 7.5). Consider an alcohol/CHG test product; the testing requirement for this configuration is four products:

1. Test alcohol/CHG product (final product)
2. Alcohol control product (effects of alcohol)
3. CHG control product (effects of CHG)
4. Product currently on the market as a control product

The regression slopes are not the same in this case. An analysis of covariance (ANCOVA) has this as a restriction that has not been met. Also, how can you instruct the reader in this more complex analysis? It is recommended, instead, that the microbial counts be *standardized* by using reductions. That is,

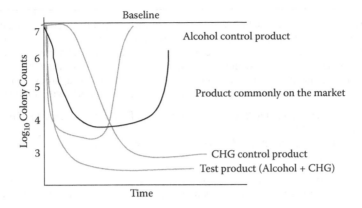

FIGURE 7.5 Comparison of product regression slopes.

Log_{10} Colony Count Reductions = Log_{10} Baseline – Log_{10} Counts at the time points.

The log_{10} colony count reduction automatically adjusts the counts as long as you take the baseline control reading of all the counts. If this is not done, the same problem will occur.

1. Studies to Replace a Component (Toweling, Wraps, etc.)

Currently, there are a number of firms that have a product such as CHG already on the market. But what happens when they want, for example, to change the towel that delivers the CHG or the outer wrapper? The comparison is the new product versus the old one, to determine equivalence; that is, the old product equals the new product.

Figure 7.6 gives an example of what the FDA looks for in both the new and old products. But what is required for the old and the new products to pass the FDA requirements? It certainly is not as direct as determining if the new product equals the old product.

For example, the former preoperative skin preparation study was to indicate that the product passed the test if it achieved an average of 3 log_{10} reductions at the inguinal site and 2 log_{10} reductions at the abdominal site. But recently, these requirements have been tightened, requiring that the lower bounds of the confidence interval $\left(\text{Lower bounds} = \bar{x} - 1.96\sqrt{\dfrac{MS_E}{n}}\right)$ are greater than 3 log_{10} reductions at the inguinal site and 2 log_{10} reductions at the abdominal site.

1. The primary analysis should not use the binomial distribution, but rather the normal distribution. Because the FDA is now tightening the requirements, do not use the two-sample t-test as listed, but instead, use a two-factor analysis of variance (ANOVA) model. The two-factor ANOVA model will evaluate 2 to n products and multiple time frames, using one variance (MS_E) for all of these factors. However, separate ANOVAs should be done for the inguinal and abdominal sites. This will allow the comparison of the different products at different times, while keeping the error rate or variance low.

The primary analysis should be based on the proportion of patient successes (responders) as a binomial endpoint. Success for a patient is defined as meeting the required 3 $\log_{10}$ reduction from baseline at the groin site and 2 $\log_{10}$ reduction at the abdomen site at 10 minutes (immediate effectiveness) and 6 hours (persistent effectiveness) after application.

In addition, adequate effectiveness of the test product and the appropriateness of the active control as an internal standard should be established based on the following:

- The lower bound of a 95% confidence interval for the responder rate of the test product ≥ 70%
- The lower bound of a 95% confidence interval for the responder rate of the active control ≥ 70%

The primary analyses should demonstrate efficacy for both dry sites (abdomen) and moist sites (groin) based on both immediate and persistent success rates demonstrating the following:

- Superiority of test vs. vehicle control based on the lower bound of a two-sided 95% confidence interval for the difference in responder rates exceeding zero.
- Superiority of active control vs. vehicle control based on the lower bound of a two-sided 95% confidence interval for the difference in responder rates exceeding zero.

There are multiplicity issues when analyzing the study with the current study design because two test materials are being evaluated for efficacy. To address these issues, you may consider using a "gatekeeper" approach to the analysis or an approach that adjusts the overall type I error rate (e.g., Bonferroni).

Alternatively, the study design can include two test materials for evaluation.

FIGURE 7.6   FDA example.

2. Computation of the lower bounds of the 95% confidence interval for these two test products (new cloth) and the old cloth are required. They are compared for equivalence. Just know that they must be equivalent, or the same. One needs the entire confidence interval (upper and lower bounds) to determine if they are equivalent. Both of the lower bounds of the confidence interval must exceed 2 and 3 $\log_{10}$ reductions.

In addition, in using the binomial distribution, there must be ≥70% of the people within the study who had greater than 2 or 3 $\log_{10}$ reductions. These people would get a "1," as they passed. Those who got below 2 and 3 $\log_{10}$ reductions would get a zero. This just makes sure there are no outliers present to pull the average values down. We will cover this in the next section.

2.   95% Responder Rate

The FDA has presented a 95% responder rate to help find a distribution that is not normal or contains outliers on the upper end.[*] This statistic makes certain that no

---

[*] For example, on the abdominal test site for the preoperative preparation, the distribution is not normal, because the FDA does not allow participants in the study who have less than 10^3 (3 $\log_{10}$/cm^2) microorganisms on that site. Instead of looking like this ⟋‾⟍ (a normal distribution), it looks like this ⌐⟍. A distribution that contains outliers appears like this: ⌐⟍ • •.

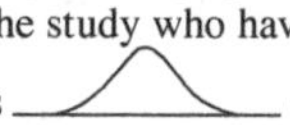

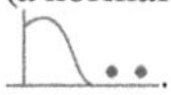

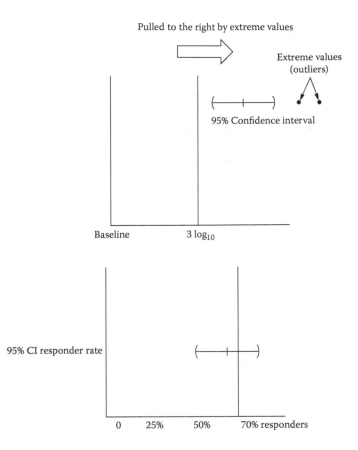

FIGURE 7.7 Responder rates.

extreme values have significantly contributed to the product passing the 95% confidence interval.

For example, if there is a product that passes the 95% confidence interval for 3 $\log_{10}$, and the data are normally distributed, the responder level will also pass the test.

If your results were based on extreme values, the confidence interval would pass, because the distribution of extreme values pulls the product to the right. However, the responder rate will fail because it is unaffected by extreme values, due to its nonparametric basis (Figure 7.7). This is a safe way of checking the data in a study, instead of using stem-leaf displays and letter-value graphs commonly used in EDA. Also, if your product does pass the lower bounds of the 95% responder rate in a pilot study, but barely, this can be a potential problem.

3. Computation
The formula for this computation is

$$\pi = p \pm z_{\alpha/2}\sqrt{\frac{p(1-p)}{n}} \pm \frac{1}{2n},$$

where:

π = 95% confidence interval of responder rate success.[*]

p = proportion of success of $\dfrac{\text{number of successes}}{\text{number of total observations}}$.[†]

n = sample size, which is the number of total observations.

$z_{\alpha/2}$ = 95% confidence interval = 1.96 for a two-tail confidence interval.

The Yates correction factor is also used, which is $\dfrac{1}{2n}$ or $\dfrac{0.5}{n}$.

This statistic works best when the probabilities are not extreme. They should be between 0.3 and 0.7.

There is an easy way to remember this:

$$np \geq 5$$

and

$$n(1-p) \geq 5.$$

When p or $1 - p$ is close to 0 or 1, the statistic should be replaced by the Poisson distribution of extreme values. We shall, however, use the standard computation, because the FDA has not mentioned anything about this situation.

For those interested, the Poisson statistic is

$$\pi = p \pm z_{\alpha/2} \sqrt{\frac{p}{n}}.$$

Do not use the Poisson statistic without contacting the FDA for guidance [100].

What if you want to check a study to see if the product reaches the $\log_{10}$ reductions that are required during the course of the study? How does one do that?

This requires a process termed an *interim analysis*. When one performs an interim analysis, one can look at the data twice or any number of times. But there is a very large cost in subject size of the study associated with this process. Suppose one were to take two looks at the data, one in the middle (interim analysis) and then one at the end (already scheduled).

The *middle look* makes the Type I error (α) change; the new α' error for the final study results is

$$\alpha' = 1 - (1-\alpha)^k,$$

---

[*] "0" = lower than the 3 $\log_{10}$ reductions; "1" = equal to or higher than the 3 $\log_{10}$ reductions. The values that received the "1" are successes.

[†] Above the $\log_{10}$ requirement of the FDA.

where:

- k = number of times the study is evaluated. Usually, this is one time, but in this case, it is two.
- α = Type I error, which is 0.05.

Using the formula

$$\alpha = 1-(1-0.05)^2,$$

α = 0.05, which is 1 study review. (Remember, for any interim analysis, one reviews the study one time [when the study is in progress] to see if the results appear favorable. Then a final review is made if the study continues.) This is 1 + 1 = 2 reviews.

$$\alpha' = 1-(1-0.05)^2$$

$$\alpha' = 1-(0.9025)$$

$\alpha' = 0.0975$ is the α for two study reviews, in this case.

To get the α to the 0.05 level, one must make the α smaller than it was before. Let us pick an α smaller than 0.05, say, $\alpha = 0.025$. That is, we tighten up the study to account for the extra look. Let us check it to make sure it is the correct one.

$$\alpha' = 1-(1-0.025)^2$$

$$\alpha' = 1-(0.9506)$$

$\alpha' = 0.0494$, which is slightly less, and no more than 0.05, so it is okay. It is conservative. One more thing that must be done is a sample size adjustment.

$$n \geq \frac{4(s^2)(z_{\alpha/2}+z_\beta)^2}{d^2},$$

where:

- 4 = number of products evaluated.
- s^2 = variance of the data. This is set at 1.00 in this example.
- $z_{\alpha/2}$ = Type I error, which changes. It was $z_{\alpha/2} = z_{0.025} = 1.96$. It is now $\alpha_{0.25/2}$ = (in the z scale) $z_{\alpha/2} = z_{0.0125} = 0.5-0.0125 = 0.4875 = z_{\alpha/2} = 2.24$, because of an extra review of the data.
- z_β = Type II error, which is 0.0842, and does not change.

d = detection level ±0.5 $\log_{10}$ does not change.
n = sample size per arm does not change.

So

$$n \geq \frac{4(1)(2.240+0.842)^2}{0.5^2},$$

or $n \geq 152$ samples for each of the four arms of the study.

We had calculated 125 subjects before we accounted for this interim analysis. There will have to be $152 - 125 = 27$ additional subjects per arm.

Be sure you do not decide to perform this procedure once the study has begun. Do it before, anticipating the need for additional subjects. They will need to be included in the correct randomization schema.

4.   Once the Data Are Examined, What Shall I Do with Them?

If you look at blinded data before the study is completed, you want to get the most use of it, so there are three things to do:

1. Look at the MS_E (mean square error) term, that is, s^2. This will tell you if the study is acceptable in terms of the error variability. The MS_E (or s^2) should be ≤1.0. To compute the MS_E, use the statistical model in the protocol. For example, $y = Blocks + A + B + (A \times B) + e$, because this will give the lowest value for $MS_E \left(= s^2\right)$.
2. Calculate the confidence intervals of the data from the products tested (Figure 7.8). I would check mainly the one that is to replace the old product (A_1 vs. A_2).

$$\text{Use } \bar{x} \pm 2.240\sqrt{\frac{MS_E}{n}}.$$

(Note that 1.96 has been replaced with 2.240 for two looks.)

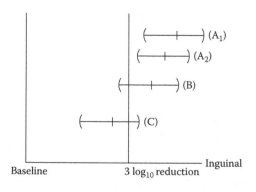

FIGURE 7.8   Calculated confidence intervals.

Looking at the inguinal site only in this example, note the following:

Because the lower bounds for both these products is ≥ 3.00 $\log_{10}$, it means that the confidence intervals (A) A_1 and A_2 products pass as they are. Everything is okay. The study can continue.

Confidence interval (B) means the product fails the test, but they can pass the study requirements by using more subjects. As long as the mean (average) is more than 3.0 $\log_{10}$ on the inguinal sites, this can be done.

Confidence interval (C) means the average is less than 3 $\log_{10}$. Stop the study. It would do no good to finish the study, because the product will not pass, and if this happens, you must go back to the drawing board. For example, you may have to apply more product on the skin site, you might have to institute a longer application period, or scrub the area harder while prepping. These should be checked using pilot studies.

3. Use Bayesian analysis to predict the study to pass at $\alpha = 0.01$. This will help you "sleep," for you are assured of its passing the test. Unlike parametric statistics that use only one test, similar to that used in the analysis of the data in this study, the Bayesian statistic uses the results from one study to predict whether a product will pass the test in a subsequent study.

8 Specific Statistical Designs for Clinical Trials

In the past, a simple t-test was the expected data analysis tool in clinical trials. It would tell if a control product and a test product were different. But as the Food and Drug Administration (FDA) keeps tightening its requirements allowing a company's product to pass a test (e.g., from a mean $\log_{10}$ reduction in microorganisms to the lower limits of the confidence interval), the statistical model should also be more stringent to allow it to be more powerful.

Let us examine the tests to see what we can do to tighten them up.

I. PREOPERATIVE SKIN PREPARATION

Because this study shares data from abdominal and inguinal sites with other products, the statistical design should be *blocked*. This will greatly reduce the error among the subjects, making the test more powerful.

For example, suppose a client has a new chlorhexidine gluconate (CHG) and alcohol product to bring to market. The clinical study test configurations will be

- CHG, by itself,
- alcohol, by itself,
- the complete test product of alcohol and CHG, and
- a product to serve as the control.

This scheme allows the products to be tested in individual components (alcohol and CHG) to make certain both products are antimicrobial by themselves. The test then compares the alcohol and CHG product in the final form. This must meet the FDA $\log_{10}$ reduction requirements. The control product is in the test to be sure that normal responses to the study are as they always have been in the past.

Four products in total must be evaluated. Two products will be applied to each subject, as shown in the figure in Table 8.1. We can see that six subjects make up one block. The six-subject block will be used for each product for three replicates. Notice that the sample size of this study must be equally divisible by 6, to be certain that it is blocked correctly, as in Table 8.2.

The sample size necessary for completing the study (first part of Chapter 7) was 94 subjects per product. That value would not be divisible evenly by 6 ($94 \div 6 = 15.666$). We see that 16 blocks of subjects would be the least we could use (Blocks × Total Block Size = Number of Replicate Subjects [$16 \times 6 = 96$]). Ninety-six subjects is the fewest number of subjects that can be used. Using

TABLE 8.1
Blocking Schema

		Blocks of Subjects		
Products	1	2	3	4

	Subjects	Products	
		A (Left)	B (Right)
	1	1	2
	2	1	3
1 Block	3	1	4
	4	2	3
	5	2	4
	6	3	4

A   B

Abdominal site
Inguinal site

R . L

R ∧ L   A = Right
B = Left

TABLE 8.2
Correct Subject Blocking

Blocks	Number of Subjects
1	6
2	12
3	18
4	24
.	.
.	.
.	.
20	120

94 replicates (subjects) per product, there are four products, and each product can be used two times on a subject (left and right sides).

$$\frac{\text{Number of subjects}}{\text{needed to complete study}} = \frac{(\text{Number of products tested})(\text{Replicates per product})}{2} = \frac{(4)(94)}{2} = 188$$

(Number of subjects
needed to finish the study
with 4 products used)

There are 188 subjects needed to finish this study to provide a sample size of 94 subjects per each of the four products. This is how subjects are blocked in this preoperative preparation study.

A.   STATISTICAL MODEL

The analysis of variance (ANOVA) model for this study is a blocked, two-factor design:

$$y = Blocks + A + B + (A \times B) + e,$$

where:

y = Log_{10} Colony Counts.
Blocks = Subjects (in blocks of 6)[*]
A = Times:
  0, if Baseline
  1, if 10 minutes
  2, if 4 hours.

There are times when the baselines are not equivalent, so using the Reduction = (Baseline − Time [10 minutes, 4 hours]) equation is more appropriate. Many people state that an analysis of covariance (ANCOVA) should be used (see Chapter 7). ANCOVA requires that the slopes of the model are the same, but this does not happen, because the four different test products have very different rates of reduction. The ANCOVA is also handier to convey results to the reader.

Instead, I suggest taking the reduction value = baseline − colony counts at 10 minutes and 4 hours.

B. ANALYSIS OF MEANS

For many applications in which the statistical results are not really understood by those reading the final report, there is a simple way to present the data, using the analysis of means (ANOM).

Let us take an example. Can the application of a product and the dry times be improved in order for the product to pass the preoperative skin preparation requirements? A pilot study was conducted with three subjects per group. The groups were type of application (cotton, gauze, or sponge applicator), and the amounts of product evaluated were 5 mL, 10 mL, and 15 mL.

The analysis of means does not require complicated statistical processes. Only the mean values and the 95% decision levels are presented. The data must be normally distributed and have standard or equal variances. These data were collected as reductions in microorganisms (see Table 8.3)

Looking at Figure 8.1, we can see that there was interaction between A (product) and B (application type) for Product 3 and the 15 mL application. All but one of the data points are between the upper and lower decision levels. It is always good to examine the mean values of the statistic, too (see Descriptive Statistics, Table 8.4). One can compute the volume and applicator types separately. Ignoring the interaction, for Product Volume (A), it shows that 5 mL is the worst, and 15 mL is the best. For Applicator Type (B), it can be seen that gauze is the worst, and the sponge applicator is the best. The best type of applicator was 3 (sponge applicator), used with 15 mL of product, as seen in Figure 8.1. One gets the best results using the sponge applicator and 15 mL of the product.

---

[*] Sometimes the blocks do not work because they become uneven as subjects and are dropped from the study (i.e., low microbial counts, missed appointments, etc.).

TABLE 8.3
Collected Data

			B (Application Type)		
			1	**2**	**3**
			Cotton	**Gauze**	**Sponge Applicator**
			2.5	2.4	2.8
	1	5 mL	2.3	2.2	2.7
			2.7	2.6	2.5
			2.7	2.6	3.0
A (Product Volume)	2	10 mL	2.8	2.8	2.9
			2.6	2.5	2.9
			2.8	2.7	3.5
	3	15 mL	2.9	2.9	3.7
			3.0	2.8	3.4

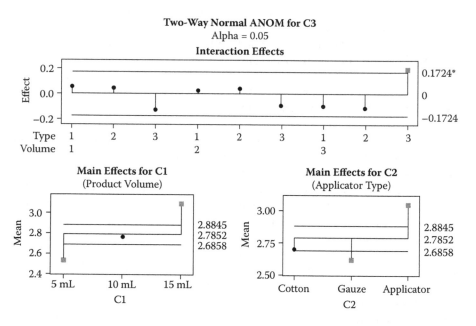

FIGURE 8.1 Analysis of means model. * Upper/lower Decision limits = $\bar{\bar{y}} \pm h\left(\alpha; I, n-I\right)$ $\sqrt{MS_E}\sqrt{\dfrac{I-1}{n}}$. Type = Applicator Type: 1 = Cotton; 2 = Gauze; 3 = Sponge Applicator. Volume = Product Volume: 1 = 5 mL; 2 = 10 mL; 3 = 15 mL.

TABLE 8.4
Descriptive Statistics

Results for C1 = 1 (5 mL application)

Variable	C2	N	Mean	SE Mean	Std Dev
Cotton	1	3	2.500	0.115	0.200
Gauze	2	3	2.400	0.115	0.200
Sponge Applicator	3	3	2.6667	0.0882	0.1528

Results for C1 = 1 (10 mL application)

Variable	C2	N	Mean	SE Mean	Std Dev
Cotton	1	3	2.7000	0.0577	0.1000
Gauze	2	3	2.6333	0.0882	0.1528
Sponge Applicator	3	3	2.9333	0.0333	0.0577

Results for C1 = 1 (15 mL application)

Variable	C2	N	Mean	SE Mean	Std Dev
Cotton	1	3	2.9000	0.0577	0.1000
Gauze	2	3	2.8000	0.0577	0.1000
Sponge Applicator	3	3	3.5333	0.0882	0.1528

9 Epistemological Requirements in Evaluating the Effects of Specific Treatments

It is important to have clearly in mind the purpose of an evaluation. Just what is it one is attempting to evaluate? For example, if a specific therapy is claimed to alleviate an allergy, it is important to define precisely what is meant by *allergy*. Does the therapy treat allergic symptomatology, the immune response effects of B-cell immunocytes secreting immunoglobulin AE, or both? Does it mean therapy in terms of clinical reduction of IgE in the bloodstream, and therefore focus on histamine reduction? Does it mean subjective relief from allergies for a patient? Perhaps both? Whatever it is, it is critical that one articulate this precisely.

I. EXPERIMENTAL STRATEGY

Once the purpose of the inquiry has been determined, one can decide how to evaluate a specific treatment modality. The data collected may be *qualitative*, such as a person's perceptions, feelings, beliefs, experiences, or goals. Or they may be *quantitative* measurements such as blood chemistry, blood gas concentrations, EEG brain wave patterns, and EKG heart function patterns. The important point is that the conclusions reached are drawn based on the experimental results, that is, the conclusions are grounded in the data. The main experimental study strategy is to assure that the conclusions reached are valid from both an internal and external perspective.

II. CHOICE OF QUALITATIVE OR QUANTITATIVE METHODS

When one considers evaluation design alternatives, the strengths and weaknesses of qualitative and quantitative data must be taken into account. Qualitative methods permit the investigator to study the clinical treatment in depth and in detail, without being constrained by predetermined scales for measurement of treatment efficacies [84]. Quantitative methods require the use of standardized measures because they focus on objective, readily observed parameters (blood pressure, EEG for brain function, EKG for heart function, etc.) [85].

An advantage of a quantitative approach is that it allows measurement of treatment responses in such a way as to provide linear data or data that can be linearized, a basic requirement of the statistical models most useful in testing quantitative data [15]. This enables the investigator to obtain a broad, generalizable set of findings that are both succinct and parsimonious. Conversely, qualitative methods typically produce much detailed information about smaller groups of patients and treatments [84]. This can provide understanding of the cases and situations studied, but at the cost of reduced generalizability.

Validity in quantitative investigations often depends crucially on careful instrument construction and operation that assure the instrument measures what it is supposed to measure. The focus is on the measuring instrument, which evaluates the effect of the treatment modality. In qualitative evaluative designs, the investigator is the instrument. Validity in qualitative studies thus depends, to a great extent, on the skill, competence, and rigor of the person doing the study.

Because qualitative and quantitative methods have different strengths and weaknesses, they offer alternative, but not mutually exclusive, strategies for evaluating various treatment modalities. For many studies, both qualitative and quantitative methods, or either, can be used to evaluate the same data.

III. STATISTICAL MODELS

The vast majority of quantitative research designs utilize biostatistical methods to evaluate the outcomes in studies that assess therapeutic modalities from a general or group perspective. It is important to select appropriate statistical models (e.g., linear regression, analysis of variance [ANOVA], analysis of covariance [ANCOVA], Student's t-test, or others) that utilize the resultant data effectively, keeping type I and II errors to a minimum. Recall that α error (type I error) is committed when one rejects a true null hypothesis, and β error (type II error) is committed by accepting a false null hypothesis [86, 87]. In other words, α error occurs when one states that there is a difference between treatments when there really is not. β error occurs when one concludes that there is no difference between treatments when there really is. The easiest way to control both α and β errors is to use more subjects (replicates) so that the possibility of both α and β errors is reduced [88]. Otherwise, merely adjusting the α error to a very small level will increase the probability of β error.

If one wishes to evaluate the efficacy of a particular treatment approach through the use of statistics, the approach must be planned carefully to assure that testing of the study data will be valid. This enables the investigator to decide on the data that are appropriate for analysis through application of a particular statistical method. Furthermore, when an investigator collects treatment data that are subject to uncontrollable experimental error, statistical methods are the only approaches that can provide valid conclusions from the data [14].

There are two components primary to any statistical evaluation: the design of the experiment and the statistical analysis of the data. These are interdependent because the method of analysis depends directly on the design employed.

There are three basic principles in statistical experimental design:

1. Replication
2. Randomization
3. Blocking

Replication means that the basic experimental measurement is made repeatedly in the same units of measure. For example, if one is measuring the CO_2 concentration of blood, the measurement would be repeated under circumstances that are experimentally the same. Replication serves several important functions. First, it allows the investigator to estimate the experimental error (standard deviation). This estimate of error becomes a basic unit of measurement for determining whether observed differences in the data are actually *statistically significant*. Second, if the sample mean (i.e., $\bar{x}$) is used to estimate the effect of a treatment factor in an evaluation, replication allows an investigator to obtain a more precise estimate of this effect. If σ^2 is the variance of the data, and there are n replicates, then the variance of the sample mean is $\sigma_0^2 = \dfrac{\sigma^2}{n}$.

The practical aspect of this is that if no replicates are made ($n = 1$), the investigator may be unable to make a satisfactory inference about the effect of a particular treatment modality. The observed difference could be experimental error and not a true difference. However, if the sample measurement were replicated more than once, and if the experimental error were relatively small, then the difference observed between treatments could be concluded to actually exist $(\bar{x}_1 \neq \bar{x}_2)$.

Randomization of testing procedures performed on subjects is the mainstay of statistical analysis [90]. No matter how objective an investigator is, experimental bias may creep into the study unless random-sampling methods are used. Often, investigators create experimental bias when they try to randomize an experiment subjectively. The best way to avoid this form of bias is through a formal scheme of random sample selection, using a table of random digits [86]. Such tables are available from various sources, including most basic statistical texts and random number–generating computer programs [86, 88, 90]. Through randomization of both treatment and control, subjects are assigned to the numbering system, and the order in which they are evaluated is randomly determined. That is, each subject is equally likely to be selected for a particular treatment at a particular time. Virtually all statistical methods require that the treatment observations (as well as their errors) are independently distributed random variables [90]. By randomization, one can average out the effects of extraneous factors (confounding variables) that are present. For example, in a study using both males and females in a treatment program, randomly assigning subjects to each treatment will average out weight differences of males versus females, which may be important in the dosage level assigned.

Blocking is a technique used to increase the precision of an experimental design [85]. A block is a portion of a treatment evaluation and it involves making a comparison of randomly assigned treatments within each block. For example, in a crossover design, one-half of the subjects are assigned Treatment A, and the remainder, Treatment B. After a "wash-out period," those who got B now get A, and vice versa. Time is the *block* variable in this case.

The exact statistical model selected depends, in part, on the data distribution (normal, skewed, bimodal, exponential, binomial, or other) [15, 85, 88, 90]. The use of exploratory data analysis (EDA) procedures can help the investigator select the appropriate statistical model and develop an intuitive feel for the data before the actual statistical analysis occurs [91].

IV. STATISTICAL MODEL-BUILDING AND ANALYSIS: EXPLORATORY PHASE

Scaled-down pilot studies are very useful in exploring the effects of a certain treatment regimen before expending time and money on a full-scale study [14]. Study objectives, experimental design, and measurement methods often can be adapted from other studies examining a similar independent (or controlled) variable. The dependent or response variable—the treatment effects—cannot be known; hence, exploration of those collected data will be very useful in selecting the appropriate statistical models to evaluate the data as inexpensively as possible. This is done through EDA. EDA is used to evaluate the data and to develop an intuitive *feel* for them [14, 91]. The process consists of four major facets:

1. *Data displays* allow the investigator to observe and intuitively comprehend the data distributions and patterns under study.
2. *Residual displays* enable the investigator to fit an appropriate statistical model to the data by interactive statistical model-building. From the residual values (the difference between the actual data values and the expected numerical data values predicted by the statistical model) as an indicator of the statistical model's predictive ability, the investigator can generate a very effective, reliable statistical model.
3. *Reexpression* often simplifies the statistical analysis by rescaling the data through any of a variety of mathematical restatements, such as reciprocals, square roots, and $\log_{10}$ restatements. When working with nonlinear data, such as those encountered in clinical trials, the reexpression procedure serves to linearize the data so a simple linear statistical model can be applied to them, instead of more complex nonlinear functions.
4. *Use of resistant and robust mathematical models* are affected little by the extreme data value outliers known to significantly affect parametric statistical models. Use of robust models often produces more reliable evaluations, even when using very few subjects, as in pilot studies.

Also, by employing EDA, one can often select a statistical model that is very representative of the real-world situation [13]. This makes the statistical analysis of the product's performance more accurate and precise, and therefore, more realistic.

To select the most optimal statistical model—the one that realistically portrays the data—several EDA data-handling procedures and displays stand out. They include the following:

1. *Stem-and-leaf* displays provide a flexible, effective procedure for ordering raw data, and thereby determining their distribution (e.g., the normal or a log-linear distribution).
2. *Letter value displays* summarize and describe raw data in terms of their dispersion relative to the median value. That is, they show where the values lie, or do not lie, relative to the median, and provide insight into the distributional shapes (e.g., skewed left or skewed right) of the data.

3. *Box plots* graphically display values in a format that resembles the standard Student *t*-confidence-level diagram. Box plot displays are strictly visual, with no numerical values provided. One can use them to compare groups of data, much like one uses a series of 95% confidence intervals (CIs).

Once a good understanding of the data distribution is obtained, a statistical model to be used with an appropriate sample scheme is chosen.

V. STATISTICAL MODEL SELECTION

The statistical model, to be appropriate, must measure the data accurately and precisely [15]. The test hypothesis should be stated as clearly and concisely as possible. If, for example, the study is designed to test whether or not products A and B are equivalent over the course of multiple washings, the statistical model should accurately test that hypothesis and measure it.

Roger H. Green [99], in his book *Sampling Design and Statistical Methods for Environmental Biologists*, describes ten steps for effective statistical analysis. These steps are applicable to any topical antimicrobial product analysis:

1. State the test hypothesis concisely to be sure that what you are testing is what you want to test.
2. Always replicate the samples. Without replication, measurements of variability may not be reliable.
3. Insofar as possible, keep the number of sample replicates equal throughout the study. This practice makes it much easier to analyze the data and produces more reliable results.
4. When determining whether a particular condition has a significant effect, be sure to take samples both where the test condition is present and where it is absent. (Example: If you find, through analysis, that reduction in a product's active ingredients neutralizes its measured effects, you should demonstrate that this problem can be corrected by increasing the level of active ingredients.)
5. Perform a small-scale study to provide a basis for sampling design and statistical model selection before going to the trouble and expense of analyzing the entire manufacturing system.
6. Verify that the sampling scheme you devise actually results in a representative measure of the population you want to evaluate. Guard against systematic and experimental bias by using techniques of random sampling.
7. Break a large-scale sampling process into smaller components.
8. Verify that the collected data meet the statistical distribution assumptions. In the days before computers were commonly used and programs were readily available, blind assumptions had to be made about distributions. Now it is relatively easy to test these assumptions, at least in part, before accepting the statistical model as valid.
9. Test your model thoroughly to make sure that it is useful for the process under study. And, even if the model is satisfactory for one set of data, be certain that it is adequate for other sets of data derived from the same process.

10. Once these nine steps have been carried out, one can accept the results of analysis with confidence. Much time, money, and effort can be saved by following these ten steps to system analysis.

Once an investigator has a general understanding of the product's attributes, s/he must fit a statistical model to the data [15, 88, 90]. At times, the data are such that nonparametric models are more appropriate than are parametric models [113]. For example, if budgetary or time constraints force the investigator to use only a few subjects per product in a study, or if some requirements of the parametric model, such as a normal (Gaussian) distribution of the data, cannot be achieved, then the nonparametric model is the model of choice.

Prior to mounting the large-scale study, the investigator should reexamine (a) the test hypothesis; (b) the choice of variables; (c) the number of replicates required to protect against type I and type II errors; (d) the order of experimentation process; (e) the randomization process; (f) the appropriateness of the statistical model used to describe the data; (g) the data collection and data-processing procedures, to ensure that they continue to be relevant to the study; and (h) the variance and standard deviation [14]. Also, once again, the product must be marketed to be successful. Therefore, another review of the product in relation to the needs of and competition in the marketplace is in order [114].

Let us now address briefly the types of parametric and nonparametric statistical models available.

VI. PARAMETRIC STATISTICS

Parametric statistics require that certain conditions be met and, generally, these include the following [15, 85, 88, 90]:

1. The observations must be independent. That is, the selection of any one case from the population for inclusion in the sample must not bias the chances of any other case for inclusion, the value which is assigned to any case must not bias the chances for any other case for inclusion, and the assigned score must not bias the score that is assigned to any other case.
2. The observations must be drawn from normally distributed populations (the so-called Gaussian distribution).
3. The populations sampled must have the same variance.
4. The variable involved must have been measured at least on an internal scale, so it is possible to perform arithmetic operations (adding, dividing, squaring, finding means, etc.).

All of the above conditions are requirements of parametric models. Ordinarily, these assumptions are not assured through statistical model checking and validation procedures. Rather, they are presumed to hold. The meaningfulness of the results of a parametric test, however, depends upon the validity of these assumptions.

Parametric statistics, which include the Student's t-test, linear regression, ANOVA, and ANCOVA, utilize parameters (mean, standard deviation, and variance). The data collected are termed *interval* data (102.915, 1×10^{-5}, 7.23914 ...). Interval data can be ranked and subdivided into an infinite number of intervals that are objectively meaningful. In other words, a scale measurement is applied that is continuous and repeatable, regardless of who takes the measurement; subjectively, insofar as divisions in the measuring device have meaning, plays no role. Usually, interval data relate to some standard physical measurement (e.g., height, weight, blood pressure, the number of deaths, etc.). Subjective perception of degrees of pregnancy, prestige, social stress, etc. cannot be translated into interval data, despite a number of research designs erroneously categorizing them as such. Extreme caution must be used when using quantitative designs to measure what is really qualitative (noninterval) data [15, 85, 88, 90].

A. ANOVA

ANOVA is the most common parametric statistic used to compare more than two groups. There are a number of variants of this model, depending upon the number and combination of groups, categories, and levels one desires to evaluate. Common forms of ANOVA include one-factor, two-factor, and three-factor designs, as well as crossover and nested designs.

VII. NONPARAMETRIC STATISTICS

A nonparametric statistic does not require the four general conditions necessary for use of parametric statistical models, with the exception of random sampling [92, 93, 113]. Nonparametric statistics do not utilize parameters (mean, standard deviation, and variance) and can be applied to evaluate noninterval data—nominal or ordinal data—but interval data as well, if appropriate.

Nominal data can be grouped, but not ranked. Data such as right/left, male/female, yes/no, and 0/1 are examples of nominal data, and such data consist of numbers used only to classify an object, person, or characteristic. Ordinal data can be both grouped and ranked. Examples include good/bad, poor/average/excellent, or lower class/middle class/upper class, which require subjective evaluation translated to a numeric scale. Nonparametric statistics also may be applied to interval data when the sample size is very small, and the data distribution cannot be assured to be *normal*, a requisite for use of parametric statistics. A *normal* (bell curve or Gaussian) distribution is not a requirement of many nonparametric models [113].

Common nonparametric models include the following:

1. *Mann–Whitney Statistic*: This test is the nonparametric analog to the Student's *t*-test and it is used to compare two groups. Unlike the parametric Student's *t*-test that assumes a normal bell-shaped distribution, the Mann–Whitney statistic requires only that the sample data collected are randomly selected.
2. *Kruskal–Wallis Model*: This is the nonparametric analog to a one-factor ANOVA model and it is used to compare multiple groups of one factor. For example, suppose one wants to evaluate the antimicrobial effects of five different hand soaps; the Kruskal–Wallis Model could be employed for this evaluation.

3. *Friedman Model*: This is the nonparametric analog to a two-factor ANOVA. It can measure any number of products, and they are blocked to reduce error.

In many therapeutic evaluations, where the number of human subjects required to perform the study is low, and thus cost-feasible, the best quantitative choice may well be the nonparametric statistical designs. Most investigators simply will not have the funding available to perform elaborate studies. Additionally, since the statistical error of nonparametric studies will tend to be type II (stating the treatment is not effective, when it really is), when effective treatments are observed, they will tend to be highly significant [92, 93].

However, there is high error with nonparametric statistics. There are two that quickly come to mind. First, it is very hard to find differences when testing more than three samples. This is because the alpha (α) level is reduced for every test one does. Additionally, the statistical model requires a larger sample size than parametric statistics. When more than three samples are evaluated, these studies can be too expensive to conduct. So plan the study throughout, accounting for the method of statistical analysis to be used.

VIII.   QUALITATIVE RESEARCH DESIGNS

Qualitative research designs generally focus on three kinds of data collection [84, 88]: (1) in-depth, open-ended interviews; (2) direct observation; and (3) written documents. The data from interviews consist of direct quotations from patients about their experiences, opinions, feelings, and knowledge. The data collected from observations consist of detailed descriptions of the patients' treatment activities and behaviors, their actions, and the full range of interpersonal interactions and organizational processes that are part of the human/patient experience. Such data may be gleaned from document analyses that yield excerpts or entire passages from clinical or program treatment records, memoranda and correspondence, official publications and reports, personal patient diaries, or open-ended, written responses to questionnaires.

The data for qualitative analysis typically come from field work [95]. During the field work, an investigator spends time in the setting under study—a treatment program, clinical setting, retreat, or wherever situations of a treatment modality that are important to a study can be observed and people interviewed. The investigator makes firsthand observations of activities and interactions, sometimes personally engaging in the treatment activities as a *patient observer*. For example, an investigator might participate in a treatment program under study as a regular member, client, or patient, talking to people about their experiences and perceptions of the treatment modality, either informally or formally. Relevant records and documents are examined. Extensive field notes are organized into readable, narrative descriptions with major themes, categories, and illustrated case examples extracted through content analysis. The findings, understandings, and insights that emerge from the field work and subsequent analysis represent the value of quantitative inquiry.

The validity and reliability of qualitative data depend, to a large degree, on the methodologies, skill, sensitivity, and integrity of the investigator [64]. Systematic and rigorous observation involves much more than just being present and looking around, and skillful interviewing involves more than just asking questions. Content analysis requires considerably more than just reading to see what is there. Generating useful and credible data findings through observation, interviewing, and content analysis requires discipline, knowledge, training, practice, creativity, and work.

Many times, data cannot be presented well in a meaningful, numerical form. This is particularly true for individual and population interior phenomenological designs, where values and meaning are the data that must then be interpreted, not observed [94]. Many of these types of designs measure subjective *quality-of-life issues*. But they are also useful where there is suspicion that a particular therapy may be beneficial for physiological treatment, but it is not known how to measure the real effects (e.g., visualization of wellness in treating chronic diseases).

Let us now explore several of these qualitative designs—the phenomenological, the heuristic, the historical, and the evaluative—in greater detail.

IX. PHENOMENOLOGICAL DESIGNS

The method involves questioning participants concerning their perceived experience of the phenomenon under study. Data are usually collected via a tape-recorded, unstructured interview where the participants report all experiences related to the phenomenon. The researcher analyzes these self-reports for essential, recurrent themes that surround the phenomenon. In these studies, the meaning imparted to the experiences is very important. Experience includes sensory, linguistic, numeric, and symbolic content. But keep in mind, it is the meaning that is of central interest to the researcher. If one experiences positive effects from a specific therapeutic modality, which may or may not provide positive *physical* therapeutic results, that modality cannot be discarded as "useless" therapy. It may not address physical well-being, but does address quality-of-life issues (emotional, mental, or spiritual), and therefore has value. The phenomenological research design provides a means of uncovering the value.

Usually, this method requires a beginning statement by the researcher to orient the study. For example, asking how a particular treatment was experienced can be a beginning statement. The interview is designed to expose recurrent themes in an exhaustive exploration of the phenomenological experience. The analysis is time-consuming and follows a preset sequence. First, tape recordings of the interviews are usually transcribed. The nonessential words are separated from those relevant to the phenomenon. *Meaning units* are highlighted in the transcript. Redundancy is reduced to meaning units and taken from collected transcripts of the participants and categorized by the researcher into related clusters or themes. Finally, these themes are reduced to cogent, succinct, one-paragraph descriptions of the phenomenon, as described by the participants. A core theme or multiple themes are the primary findings of this study.

A potential weakness of this method lies with researcher bias of the study. Grouping meaning units, omitting material from the study, and composing descriptions of thematic clusters are all potential sources of bias. Hence, it is important to assure external validity of the study by comparing results with those of others.

X. HEURISTIC DESIGNS

This research design is used to *discover* the meaning or nature of an experience or therapeutic methods and procedures for further exploration. Discovery is the main focus. This research design uses a series of iterations to ever deepen the exploration process. The results of the first iteration are used as the starting point for the next. It is a common method used by conventional medical practitioners to diagnose disease, as well as to treat it. An example of this research approach may be seen in the treatment of an infection. A practitioner first prescribes treatment A. The results noted from the prescription are the starting point for the next treatment regimen. If no improvement is noted, the practitioner may increase the dose or switch to another regimen and then evaluate its results.

Heuristic designs involve active participation of both the researcher and subject. The researcher continually probes the subject concerning his/her experience. Probing interviews are common, using a structured or semistructured interview style. A heuristic investigation will often employ a defined set of questions, based upon a decision-tree diagram. The process of data collection is interactive and proceeds in a sequential manner.

When used as a research design, the researcher begins by asking questions concerning the phenomenon of interest. The answer generates more questions. These questions, in turn, become a catalyst for more extensive inquiry. In research, the investigator questions each of the subjects. The interview direction, of course, depends upon the subject's responses. The investigator compiles a schematic list detailing the various responses of each subject, and then compares and contrasts the responses, generating categories, clusters, and themes from the interviews.

The same general biasing phenomena noted in the phenomenological research study affect the internal validity of this kind of study. To assure external validity, it is important that the research be repeated by other investigators at different geographical locations.

XI. HISTORICAL RESEARCH DESIGN

The historical design enables the investigator to perform an evaluation using data collected in the past. The purpose of the historical method is to uncover and organize relevant data from a selected period of time, enabling the researcher to evaluate those data with relevancy to the present. It can be used for interior or exterior phenomena. In this design, the researcher compiles and systematically reviews the documentation of people, places, or events concerning the therapeutic modality under investigation.

For example, if certain individuals are able to avoid the various symptoms of AIDS indefinitely, even though they have tested HIV-positive, an investigator may explore how they have done it. This would involve exhaustive documentation of the many variables, individual or environmental, that could influence progression to active disease. Hence, success in using the design requires careful collection and organization of data, as well as *reliable data*. Historical research designs are often criticized for being subject to high degrees of biasing—from both the investigator and the historical documents used as data. Because the data were collected in the past, there is no way to assure their internal validity. The same precautions listed for use of phenomenological design also apply to the internal validity of this design. To assure external validity, it is important that the research be repeated by different investigators at different geographical locations.

XII.   EVALUATIVE RESEARCH DESIGN

This design is used to evaluate different therapeutic modalities for their efficacy in treating a disease *in the field*. The research design can either be formative (assessing a therapeutic modality being developed) or summative (assessing a fully developed therapeutic modality). This method requires that the investigator state the therapeutic goal, collect data while the actual therapy is in process, and compare the results of the treatment modality relative to its therapeutic goal. Usually, numerical data are not collected, but rather the qualitative findings of the investigator. An example of this type of research design would be determining the therapeutic benefits of "play" in cases of mild, but chronic, depression.

The evaluative research method enables the researcher to obtain reliable data, so that a decision regarding the initiation, continuation, or termination of a therapeutic modality can be made. The research must be guided by clear formulation of the program goals and the objectives of the evaluation. To assure both internal and external validity, the requirements noted in the phenomenological research design also apply to this design.

XIII.   CONCLUSION

It is important that products (or therapeutic regimens) be evaluated for merit and potential problems. There are numerous ways to perform these evaluations, using either quantitative or qualitative research designs, and it is important that investigators be familiar with a large selection of these. This will prevent the researcher, who has only one tool—a hammer—from viewing everything as a nail.

10 Evaluation Strategies and Sample Working Protocols

In this chapter, we will limit the focus to three general areas:

1. Consumer antimicrobial products
   a. Hand soaps
   b. Body soaps
2. Food-handler antimicrobial products
3. Medical/healthcare industry antimicrobial products

I. CONSUMER ANTIMICROBIAL PRODUCTS

This category consists of antimicrobial body soaps, hand soaps, towelette hand wipes, and other associated products. The purpose of these types of products is to reduce transient, potentially pathogenic microorganisms picked up in the environment (e.g., from countertops, coins, door handles, etc.). These types of products are often intended, too, to remove excess normal microbial flora found particularly in moist body areas [30]. The types of clinical trials currently used in evaluating these products include the Healthcare Personnel Handwash, the Modified Cade Handwash, the general-use handwash for hand soaps, and axilla cup scrub sampling for body soaps [20].

The Healthcare Personnel Handwash evaluation, as presented in the Federal Register, is rarely used in its entirety for evaluating consumer hand-cleansing antimicrobial products. The Modified Cade and the general-use procedures are the most commonly used. For antimicrobial bodywash products, the axilla cup scrub and the full bodywash evaluations are used.

Currently, there are several types of other methods brought about by American Society for Testing and Materials (ASTM), but the Food and Drug Administration (FDA) still holds its view on healthcare personnel handwashes.

Let us begin with the procedures most often used for evaluations of antimicrobial handwash products.

A. EVALUATING CONSUMER ANTIMICROBIAL HANDWASH PRODUCTS

Manufacturers and suppliers of consumer antimicrobial hand soaps must evaluate the antimicrobial properties of their products not only to collect in-house product information, but also to assure that a product performs as expected—that is, the product is antimicrobially effective and not excessively irritating to the hands. There are a variety of approaches available for evaluating these products, but in this book, we focus on antimicrobial efficacy evaluations employing human volunteer subjects, and on two common in vitro evaluations, the time-kill and minimum inhibitory concentration tests.

B.   EVALUATIVE CONFUSION

Methodologies for human clinical trials used to evaluate consumer antimicrobial soaps are ill-defined. This is due mainly to the FDA's decision not to provide guidelines in testing these products because it is not convinced that consumers are provided any substantial benefits by using antimicrobial soap products on a routine basis. If guidelines were provided, then by implication, it would be assumed that the FDA perceived merit in such products. At the time of this writing (second edition), this is still true.

While most involved agree that the healthcare personnel handwash evaluation is not appropriate for evaluating consumer antimicrobial soaps, some regulatory groups suggest that a *conflict of interest* is implicit in the alternatives proposed by the manufacturers. That is, the soap manufacturing associations do not have an unbiased perspective on performance of these products, a crucial requisite to designing valid testing methods.

C.   ANTIMICROBIAL HAND SOAPS AND DISEASE PREVENTION

Let us step back a moment and discuss aspects of infectious diseases and the role antimicrobial hand soaps play in their prevention. For infectious diseases to occur, the following sequence of five events must take place [12, 28]:

1. A person must come into contact with the microorganisms.
2. The microorganisms must enter the person.
3. The microorganisms must spread from the entry site.
4. The microorganisms must be able to multiply within the person.
5. As a result of microbial enzymes and toxins, and to some extent response, tissue damage occurs.

An effective handwash disrupts the disease process after event 1 by removing contaminating microorganisms from the hand surfaces. While not all of the microorganisms can be removed, their population numbers can be reduced below the level required to cause disease in normal humans [12].

When the anti-infective properties of antimicrobial hand soaps are discussed, it is important to understand what types of microorganisms must be addressed. The microorganisms that normally colonize hand surfaces pose little threat of infectious disease and, in fact, serve competitively to exclude colonization by transients [14]. There are, of course, situations where normal or resident microorganisms cause disease when they are introduced into areas where they are not *normal*. An infected cut often is an example of this. However, even in this case, appropriate washing serves to degerm the infected area, cleaning it of dead cells and exudative material.

In the vast majority of cases, the threat of infectious disease is from *transient*, pathogenic microorganisms that contaminate and may temporarily colonize hand surfaces. Hand contamination with transients occurs when one contacts substances such as mucous, blood, soil, urine, feces, or food. Given the opportunity, the contaminant microorganisms may then infect the person or be passed on to others via hand contact [20].

D. FUNCTIONAL PARAMETERS OF EFFECTIVE ANTIMICROBIAL SOAPS

Two parameters are of primary interest in evaluating antimicrobial hand soaps: their immediate degerming effectiveness and their persistent antimicrobial effectiveness [29]. Recall that immediate antimicrobial efficacy is the measure of a handwash product's effectiveness as a function of both the mechanical removal of contaminating microorganisms in the process of the handwash procedure and the immediate inactivation of microorganisms resulting from contact with the antimicrobial ingredient(s) in the soap. The persistent antimicrobial effectiveness is the product's ability to prevent, either by microbial inhibition or lethality, transient microbial recolonization of the skin surfaces after handwashing.

Accurate and precise measurement of these two parameters is often difficult. It is important, then, that efficacy evaluations be well defined and clearly stated before conducting them to assure that the *question* asked will actually be answered with valid data.

II. ANTIMICROBIAL EFFICACY STUDIES

To determine which microorganism species are susceptible to the antimicrobial soap, as well as the rates of microbial inactivation, certain in vitro tests should be conducted [14]. These include time-kill evaluations, minimum inhibitory concentration evaluations, and microbial sensitivity tests. But in order to determine how effective the handwash is in the *real world*, in vivo human product-use studies—clinical studies—must be conducted [96].

As noted earlier, study designs routinely utilized for this include the following:

1. Healthcare Personnel Handwash Evaluation
2. Modified Cade Handwash Procedure
3. General-Use Handwash Evaluation

Let us discuss these tests in greater detail. Then, I shall propose an approach to evaluation that I believe is optimal for testing the efficacy of antimicrobial handwash products.

A. HEALTHCARE PERSONNEL HANDWASH EVALUATIONS

This testing approach is utilized when evaluating antimicrobial products used in the medical field by healthcare personnel. Enough subjects are recruited to assure that the sample size is adequate in terms of reducing the probability of type I and II errors appropriate to the statistical models selected, thereby providing a high degree of statistical power (the *power of the test*).[*]

A test product, a vehicle product (test product without the active antimicrobial), and a reference product are customarily used in this evaluation. It is expected that the test product will demonstrate significant antimicrobial properties through elimination of microorganisms by mechanical removal and by microbial inactivation as a function of exposure to the active antimicrobial ingredient [1]. The vehicle is expected to demonstrate the product's ability to remove microorganisms through the mechanical action of the handwash only. The reference product is used to validate the study in that results of using the reference product are expected to be similar to those from other studies in which it was used.

A *washout period* of at least 7 days is necessary, during which time the subjects are not permitted to use antimicrobial products or expose their hands to acids, bases, or any other substances known to affect microbial populations.

On the test day, subjects are inoculated with *Serratia marcescens*, a marker microorganism strain that produces red colonies when plated on soybean-casein digest agar. This allows them to be distinguished from microorganisms normally found on the skin, which will appear white or yellowish on the agar. The hands are next sampled using the glove juice procedure to establish a baseline measurement of the number of bacteria inoculated.

The hands are again inoculated with *S. marcescens*, and then a handwash is performed, using the assigned test configuration—test product, vehicle, or control product. Following this, the glove juice sampling procedure is again employed.

The inoculation/wash procedure is repeated 10 consecutive times. Glove juice samples are taken after inoculation/wash cycles 1, 3, 7, and 10, and the dilutions plated on soybean-casein digest agar. The agar plates are then incubated at $25 \pm 2°C$ for approximately 48 hours.

The microbial population counts on the hands after washes 1, 3, 7, and 10 are then compared to the baseline average value. To be acceptable as a healthcare personnel handwash, the product must demonstrate at least a two $\log_{10}$ reduction from baseline after the first wash (immediate efficacy) and at least a three $\log_{10}$ reduction after wash 10 (persistence efficacy).

The FDA is tightening their methods by evermore requiring the product to pass the two and three $\log_{10}$ reductions, not by the average, but by the lower bounds of the 95% confidence interval.

---

[*] Type I (α) error is committed when one rejects a true null hypothesis and type II (β) error is committed by accepting a false null hypothesis. The statistical power is the ability of the test statistic to detect a true alternative hypothesis. In other words, type I error occurs when one states that there is a difference between products when there really is none. Type II error occurs when one concludes that no difference exists between products when there really is one. The *power of the test* is the ability of the test statistic to conclude that a difference exists when one really does. The power of the test is very sensitive to the amount of variance that is implicit in the data, the discreteness of the difference sought between the systems evaluated, and the size of the sample population. The smaller the variance, the smaller the difference sought, and/or the larger the sample population, the greater the power of the test.

Note: In evaluating this study, use only (1) the baseline, (2) wash 1, and (3) wash 10. Do not use the other washes (3 and 7) in the statistical model for the FDA. These data points will only cause more variability, and there is no requirement for them.

B.   MODIFIED CADE HANDWASH PROCEDURE

The number of test subjects recruited varies, but it is common to enroll 55 to 65 subjects. A washout period of at least 7 days is observed employing the same product-use restrictions presented for the healthcare personnel handwash procedure. This is followed by a baseline measurement period.

On the first day of the baseline period, subjects wash their hands five consecutive times with a bland soap. Microbial samples are collected from washes 1 and 5, using a *basin wash* procedure. The basin wash procedure requires subjects to wash their hands with the nonmedicated bland soap in a polyethylene container such as a Ziploc® freezer storage bag containing one liter of sterile water. After washing, the wash water is mixed well, and sample aliquots are plated on soybean-casein digest agar. The agar plates are incubated at $30 \pm 2°C$ for approximately 48 hours.

For days 2 and 3 of the baseline period, the subjects wash their hands ad libitum with the bland soap outside the laboratory. On day 4, those 45 to 50 subjects having the highest baseline bacterial colony counts from day 1 remain in the study for completion of day 5; the others are dismissed.

On day 5, subjects return to the laboratory to wash their hands as on day 1, and wash 1 and 5 samples are again collected as previously described. Subjects then continue using the bland soap for all handwashing outside the laboratory on days 6 and 7. Finally, on day 8, subjects return to the laboratory and again wash their hands five consecutive times, with samples collected from washes 1 and 5 using the basin wash procedure.

The subjects are then assigned their test handwash soap products. Subjects wash their hands with the product a minimum of three times per day outside the laboratory, with at least 1 hour between washes. They also use the assigned product for any additional handwashes, as well as for bathing and showering. This regimen is followed for 11 consecutive days. On day 12, subjects return to the laboratory and turn in their assigned test products. They then wash five consecutive times with a bland soap. Again, the samples are collected using the sterile basin wash procedure for washes 1 and 5.

Usually, the first wash samples collected from the baseline period are pooled, as are the samples from the fifth wash. That this can be done, however, must be verified by proving that the data to be pooled are, among them, not statistically different. The count data from the first test sample wash are then compared to the pooled wash 1 baseline data, and the count data from the fifth test wash are compared to the pooled wash 5 baseline data.

C.   GENERAL-USE HANDWASH EVALUATION

Normally, 10 to 20 subjects are recruited for each test product evaluated. A control product is sometimes included as well. A washout period of at least 7 days is observed, during which time subjects are not permitted to use antimicrobial products or expose their hands to compounds known to have antimicrobial properties [20].

The remaining study procedures are identical to those of healthcare personnel handwash, except that the total number of washes is at least 5 but as many as 10, with glove juice samples collected at baseline, washes 1 and 5, and at wash 10, if 10 washes are used. The microbial colony counts taken from samples after the test washes are compared to the baseline values.

D. PROPOSED ANTIMICROBIAL HAND SOAP EVALUATION

The optimal evaluation design should be practical, yet provide accurate and reliable results, ideally, for reductions in transient microorganisms, not resident ones [22]. The healthcare personnel handwash is designed to evaluate the antimicrobial efficacy of products used in the medical healthcare field, and therefore is excessively stringent for evaluating consumer-use antimicrobial soap products.

Results generated by the Modified Cade Handwash procedure are sometimes considered equivocal, particularly because the antimicrobial efficacy is evaluated in terms of normal microbial flora data, which is of questionable value for evaluating consumer antimicrobial hand soaps. Additionally, it measures the residual antimicrobial effectiveness of the test antimicrobial product, not the immediate effects.* Finally, because a control product is rarely used, the validity of any such study is open to question.

The optimal design for evaluating consumer antimicrobial hand soaps is a combination of the healthcare personnel handwash and the general-use handwash evaluations. It is based on the removal of transient microorganisms over the course of five consecutive inoculation/wash cycles and is practical, accurate, and reliable. The proposed study design follows.

1. Test Material

The products to be evaluated are

    Test product:
        Lot Number: _____
        Expiration Date: _____

    Control product:
        (test product without antimicrobial compound)
        Lot Number: _____
        Expiration Date: _____

_____

*

2. Test Methods

a. Subjects

A sufficient number of overtly healthy subjects over the age of 18, but under the age of 70, are admitted into the study to ensure that at least 15 subjects complete the study per test configuration group. The number of subjects used should be set, depending upon the nature of the study. For example, if the lower 95% confidence interval is used, and the test product achieves only a 3.1205 $\log_{10}$ reduction in microorganisms at the tenth wash, additional subjects will have to be used. Utilize the statistical method properly to estimate the number of subjects necessary for testing. Subjects should be of mixed age and gender, and free from clinically evident dermatoses or injuries to the hands or forearms. All subjects must sign Informed Consent Forms prior to participating in the study.

b. Concurrent Treatment

No subject can be admitted into the study who is using topical or systemic antimicrobials, or any other medication known to affect the normal microbial flora of the skin.

c. Pretest Period

A 7-day period prior to the test portion of the study will constitute the pretest period. During this time, subjects must avoid the use of medicated soaps, lotions, deodorants, and shampoos, except as supplied in a personal hygiene kit, as well as avoid skin contact with solvents, detergents, acids, and bases.* They must also be instructed to avoid contact with specific kinds of products on a restricted list that is supplied to them (include here any commercial products they will not be allowed to use). Further, subjects must avoid the use of ultraviolet (UV) tanning beds and bathing in pools or hot tubs containing chlorine or other biocides. This regimen will allow for stabilization of the normal microbial populations residing on the hands.

Before the initiation of the study, a Study Description and Informed Consent Forms must be provided to the subjects. The study must not begin until institutional review board (IRB) approval has been obtained for all facets of testing and all relevant material. Trained laboratory personnel should be available to answer any questions that may arise.

d. Neutralization

A neutralization assay must be performed to assure that the neutralizers employed in the evaluation effectively neutralize the antimicrobial activity of the product.

e. Experimental Period

Testing will comprise a period of 7 days. Each subject will be employed for about 2 hours on only 1 day of that period. A practice wash using a nonantimicrobial bland soap and the same application procedure as prescribed for the test product should

---

* Subjects should be supplied a personal hygiene kit containing a bland soap, shampoo, deodorant, and lotion, as well as a pair of rubber gloves. The gloves should be worn when exposure to antimicrobials is unavoidable. Subjects should be instructed to use exclusively the contents of this kit for personal hygiene needs during their participation in the study.

precede the actual test portion of this study. It ensures that the subjects understand the wash procedure.

After the practice wash has been completed, a 5.0 mL aliquot of a suspension of at least 1.0×10^8 CFU/mL *Serratia marcescens* (ATCC #14756, red-pigmented strain) is transferred into each subject's cupped hands. The inoculum is then distributed evenly over both hands and up to the wrist, via gentle, continuous massage for 45 seconds. After a 2-minute air-dry, the glove juice sampling procedure is performed to provide the baseline data. It is followed by a 30-second nonmedicated soap handwash. The temperature of the water used for this and all subsequent wash cycles should be maintained and controlled at $40 \pm 2°C$.

The microbial inoculum is again distributed evenly over both hands up to the wrist, via gentle continuous massage for 45 seconds. After a 2-minute air-dry, the subjects wash with their assigned product configuration according to the use directions supplied. This is followed by the glove juice sampling procedure.

This inoculation/wash procedure is repeated five consecutive times, with a minimum of 5 minutes and a maximum 15 minutes between applications. A transient microorganism sampling of the hands using the glove juice sampling procedure is performed after inoculation/product wash cycles 1 and 5.

f. Glove Juice Sampling Procedure

Following the prescribed wash and rinse, excess water is shaken from the hands and powder-free sterile gloves are put on the subjects' hands. At the designated sampling times, 75.0 mL of Sterile Stripping Suspending Fluid without product neutralizers is instilled into each glove. The wrist is secured and an attendant massages the hand through the glove in a standardized manner for 1 minute. Aliquots of the glove juice (dilution 10^0) are then removed and serially diluted in Butterfield's phosphate-buffered solution containing appropriate neutralizers.

Triplicate spread plates are prepared from each of these dilutions, using soybean-casein digest agar with neutralizers. The plates are incubated at $25 \pm 2°C$ for approximately 48 hours. Those plates providing colony counts between 25 and 250 should be preferentially utilized for enumeration. If no agar plates provide counts in the 25 to 250 range, those plates having counts closest to that range should be used in determining the number of viable microorganisms.

Following the final product application cycle (application 5) and glove juice procedure, the subjects should be required to wash with 70% isopropyl alcohol for 1 minute, followed by air-dry, and a thorough rinse under tap water in order to eliminate any remaining *Serratia marcescens* from the hands. If desired, a wash using a surgical scrub product may be added as well.

g. Methods of Analysis

The plate-count data collected from this study should be evaluated using standard statistical computer software. The $\log_{10}$ number of viable microorganisms recovered from each hand is designated the recovered or R value. It is the adjusted average $\log_{10}$ colony count measurement for each subject at each sampling time. Each R value is determined using the following formula:

$$R = \log_{10}\left[75 \times C_i \times 10^{-D}\right],$$

where:

75 = the amount of stripping solution (mL) instilled into each glove to perform the glove juice sampling procedure.

C_i = the arithmetic average colony counts for the three plates for each subject at a particular dilution level.

D = the dilution factor.

Note: A $\log_{10}$ transformation must be performed on the average plate counts to convert the data to a linear scale. A linear scale, more appropriately a $\log_{10}$ linear scale, is a requirement of the statistical models to be used.

h. Statistical Analysis   A pre–post experimental design is utilized to evaluate and compare the antimicrobial effectiveness of the test product.

Pre-Product Application	Post-Product Application
$R\ A\ O_{BL}$	$O_1\ O_5$
$R\ C\ O_{BL}$	$O_1\ O_5$

where:

R = Subjects randomly assigned to the study.
A = Independent variable: test product.
C = Independent variable: control product.
O_i = Dependent variable: microbial counts at baseline (BL) and after product-use washes 1 and 5.

Exploratory data analysis should be performed on the data to assure they meet the requirements of the statistical models used. Stem-leaf ordering, letter value plots, and box plots should be generated to assure the data collected approximate a normal (Gaussian) distribution [91]. Use a two-factor analysis of variance (ANOVA) with interaction model at the 0.05 level of significance for type I (α) error. One factor is products, and the other is sample times.

3. Comments

It is suggested that the FDA's over-the-counter (OTC) monograph be followed for the various technical nuances of the study (21 CFR Parts 333 and 369, part III; 17 June 94). A minimum of 15 subjects per group is recommended, as is the use of a control product. If one test product is used, then a minimum of 30 to 40 subjects should be employed (15 to 20 for the test product and 15 to 20 for the control product). This is a statistically adequate number, in most cases, to detect true differences between the test and control products, and between the microbial reduction counts from baseline after washes 1 and 5. But this is true only if the standard deviation value for baseline counts is less than 0.5 logs. If the standard deviation is greater, it will be difficult to detect a *true difference* between the test and control products over the two wash samples, and more difficult yet to detect *true microbial reductions* from the baseline values. Hence, using a larger sample size is in the best interest of the product manufacturers, but must, of course, be balanced against cost. However, if more products are used, then the sample size must be expanded.

If one must meet the lower bounds of the 95% confidence interval, and the test product passes the average test, then the number of subjects used will depend upon how close to the 2.00 and 3.00 $\log_{10}$ mark the product must be.

This design for general-use handwash products permits measurement of the immediate degerming effects (first wash) as well as the effects after five consecutive inoculation/wash cycles to assure that there is no cumulative buildup of microorganisms after wash sample 5. Additional inoculations/wash cycles and samples will likely add no predictive value to the evaluation.

For this evaluation, the alpha (α) level should be set at 0.05. For screening tests for general product effectiveness, the total number of subjects may be reduced to as few as 10 per group, and the alpha (α) level set at 0.10.

Prior to beginning any test procedures, a product neutralizer evaluation must be performed to assure that the active antimicrobial ingredients in the product can be neutralized and to assure that diluents used in the assay have no significant effect on *S. marcescens*. An ANOVA design should be used to analyze the data from at least four configurations:

Configuration 1 (negative control): *S. marcescens* plated on soybean-casein digest agar

Configuration 2 (positive control): *S. marcescens* and test product plated on soybean-casein digest agar (determines if product kills)

Configuration 3: *S. marcescens* and test product and neutralizer plated on soybean-casein digest agar (determines if product has been neutralized)

Configuration 4: *S. marcescens* and neutralizer plated on soybean-casein digest agar (determines if the neutralizer affects the growth of *S. marcescens*)

Plate count data from configurations 1, 3, and 4 should be equivalent at the 0.05 level of significance for the neutralizer to be considered adequate.

4. Conclusion

It is necessary that manufacturers of antimicrobial soap products accurately and precisely evaluate and measure the antimicrobial efficacy of their products under simulated in situ wash conditions employing human subjects.

III. SAMPLE PROTOCOL: HEALTHCARE PERSONNEL HANDWASH EVALUATIONS

The following are sample protocols for commonly used evaluations of consumer antimicrobial soap products. These may be modified to suit the needs of the user.

A.   MODIFIED CADE HANDWASH EVALUATION

1. Purpose

The purpose of this study is to determine the effectiveness of one test antimicrobial soap product in reducing the level of aerobic microbial flora on the hands following an extended period of use, using a Modified Cade handwashing procedure.

2. Scope

The antimicrobial effectiveness of one test antimicrobial soap product will be determined. The study will begin with 65 human subjects undergoing baseline determination sampling. Those 50 subjects having the highest baseline microbial populations will be utilized in the test portion of the study. After 11 consecutive days of using the test product, the hands will be sampled on day 12, and the microbial reductions from baseline calculated.

3. Test Material

The product to be evaluated and the placebo (nonantimicrobial) soap are:

    Test Product
        Lot Number: _____
        Manufacture Date: __ _____

    Placebo Product
        Lot Number: _____
        Manufacture Date: _____

4. Test Methods

a. Subjects

A sufficient number of overtly healthy subjects at least 18 years, but under 70 years of age will be admitted into the study to ensure that 65 subjects undergo baseline sampling. Those 50 subjects with the highest baseline microbial counts will continue into the test portion of the study. Subjects will be of mixed sex and age. On the day of the test, subjects will be free from clinically evident dermatoses or serious injuries to the hands and forearms. All subjects will have the study protocol explained to them and will sign the Informed Consent Forms prior to participating in the study.

b. Concurrent Treatment

No subject will be admitted into the study who is currently using topical or systemic antimicrobials, or any other medication known to affect the normal microbial flora of the skin.

c. Pretest Period

The 7 days prior to the test portion of the study will constitute the pretest period. During this time, subjects will avoid the use of medicated soaps, lotions, deodorants, and shampoos, as well as skin contact with solvents, detergents, acids, and bases. Subjects will be supplied a personal hygiene kit containing nonmedicated shampoo, deodorant, soap, and lotion, as well as latex gloves. Subjects must use only the products supplied in the kit during their participation in the study. They must avoid contact with products on the restricted list. If contact with restricted products is unavoidable, latex gloves will be worn. Subjects will also avoid using UV tanning beds and bathing in pools or hot tubs containing chlorine or other biocides. This regimen will allow for stabilization of the normal microbial populations residing on the hands.

Before initiation of the study, a study description will be given to subjects (patient information). Informed Consent Forms and any other supportive material relevant to the safety of the subjects will be supplied by the principal investigator, following their review and approval by an IRB. The primary purpose of the IRB is the protection of the rights and welfare of the subjects involved in a clinical study (reference CFR 21, Part 56). Trained laboratory personnel will explain the study to each participant and will be available to answer any questions that may arise.

d. Baseline Period

Fifty-five (55) to 65 human subjects will be evaluated during the baseline period. The subjects will come into the laboratory on day 1 and wash their hands five consecutive times with the sponsor-supplied placebo soap. The laboratory sink method will be used for washes 2, 3, and 4. For washes 1 and 5, the subjects will employ the sterile basin wash technique. Samples of the rinsate in the basins will be plated in triplicate onto soybean-casein digest agar plus neutralizers to determine the microbial populations.

On days 2 and 3, subjects will continue to wash outside the laboratory as usual, using only the supplied placebo soap for all normal handwashing and bathing. On day 4, the 50 subjects with the highest microbial counts from their hands will continue into the test portion of the study. The other 15 subjects will be dismissed from the study. All subjects remaining in the study will continue to use the placebo soap for all washing on day 4.

On day 5, subjects will return to the laboratory and again wash their hands five consecutive times using the supplied placebo soap. The laboratory sink method will be used for washes 2, 3, and 4. For washes 1 and 5, the subjects will employ the sterile basin wash technique. Samples of the rinsate in the basins will be plated in triplicate onto soybean-casein digest agar plus neutralizers to determine the microbial populations.

On days 6 and 7, subjects will continue to only use the supplied placebo soap for all normal washing, as needed.

e. Test Period

On the first day of the test period, the subjects will again return to the laboratory and wash their hands five consecutive times with the supplied placebo soap. The laboratory skin method will be used for washes 2, 3, and 4. For washes 1 and 5, the subjects will employ the sterile basin wash technique. Samples of the rinsate in the basins will be plated in triplicate on soybean-casein digest agar plus neutralizers to determine the microbial populations.

The subjects will then perform supervised handwashes with their test product at the laboratory starting 30 minutes after the fifth placebo-soap wash. They will wash three times with the test product with at least 1 hour apart between washes. After these washes have been completed, the 50 subjects will be issued test product samples, Laboratory and Home Handwash Log Forms, and product-use instructions.

The subjects will wash three times daily at home, each wash no less than 1 hour apart for the test period day 2, and continue washing three times daily for the next 9 days (test days 3 through 11). These washes will be documented on the Laboratory and Home Wash Log Forms. Subjects will also use their test product for any additional handwashes, and for bathing and showering.

On evaluation day 12, the subjects will return to the laboratory, turn in their issued products, the Laboratory and Home Handwash Log Forms, and perform five consecutive handwashes using the supplied placebo soap. The laboratory sink method will be used for washes 2, 3, and 4. For washes 1 and 5, the subjects will employ the sterile basin wash technique. The rinsate in the basins will be plated in triplicate on soybean-casein digest agar plus neutralizers to determine the microbial populations.

f. Sterile Basin Wash Technique

Before beginning the wash, the subjects will remove all jewelry from the hands and forearms and wet their hands in a 3.79 L Ziploc storage bag containing approximately 1.0 L of sterile deionized water. The water will be at room temperature (20 to 25°C).

Subjects will dispense two pumps of liquid placebo (or test) product and carefully rub it over all hand surfaces. After the liquid soap is spread, they then lather for 30 seconds. Subjects will wash their hands (to the wrists only) and fingernail areas by working the lather over them for 30 seconds and then rinse their hands in the bag up to the wrists for only 30 seconds.

The 1.0 L volume of rinsate will be mixed well. A 5.0 mL aliquot will be removed from the bag with a sterile pipette and diluted in 0.1 M Sterile Stripping Suspending Fluid containing appropriate product neutralizers. The diluted aliquots will be pour-plated with soybean-casein digest agar plus neutralizers and incubated at 30 ± 2°C for up to 48 hours.

g. Laboratory Sink Method

Before beginning the wash, subjects will remove all jewelry from the hands and forearms and moisten their hands under running tap water (40 ± 2°C). Subjects will dispense two pumps of liquid placebo (or test) product and carefully rub it over all hand surfaces.

After the liquid soap is spread, subjects will wash their hands, the fingernail area, and two thirds of the forearms by working up a lather over them for 30 seconds. Subjects will rinse the hands and two-thirds of the forearms in running water for 30 seconds (40 ± 2°C).

h. Incubation

All plates will be incubated for up to 48 hours at 30 ± 2°C. After incubation, the colonies on those plates that are countable will be enumerated.

i. Methods of Analysis

The plate count data collected from the study will be evaluated using the MiniTab7 (or other) statistical computer software. The estimated $\log_{10}$ number of viable micro-organisms recovered from both hand samples will be designated the *R* value. It is the adjusted average $\log_{10}$ colony-count measurement for each subject at each sampling time. Each *R* value will be determined using the following formula:

$$R = \log_{10}\left[1000 \times C_i \times 10^{-D}\right],$$

where:

    1000 = the number of mL of rinsate solution instilled into each Ziploc storage freezer bag (Sterile Basin).

    C_i = the arithmetic average colony count of the three plate counts for each subject at a particular dilution level.

    D = the dilution factor of the rinsate.

j. Calculation of Microbial Reductions

Microbial counts (microbial populations recovered from both hands) per liter from the first and fifth basin for each subject on each sampling day are converted to $\log_{10}$ scale in order to linearize the data, a requirement of the statistical models to be used. Data from both the first and fifth sterile basin wash technique samples for the three baseline days will be averaged for each individual. Geometric means of the data will be calculated as follows:

$$\text{Geometric mean} = \text{antilog}_{10}\frac{1}{}\sum \log_{10}x = \left(x_1 \cdot x_2 \cdot x_n\right)^{1/n}.$$

In this case, because the $\log_{10}$ scale will have already been achieved from formula 1, formula 2 above will be modified as represented in formula 3.

$$\text{Geometric mean} = \text{antilog}_{10}\frac{1}{}\sum R.$$

The following percent reductions will be calculated for both the test and baseline samples (using 50 test product subjects) collected after washes 1 and 5:

    A = Geometric mean of first basin wash counts on test day (day 12)

    B = Geometric mean of fifth basin wash counts on test day (day 12)

    C = Geometric mean of three baseline basin wash counts for first wash

    D = Geometric mean of three baseline basin wash counts for fifth wash

Percent microbial reduction day 12, first basin wash compared to the pooled baseline samples from first washes:

$$\text{First wash} = \frac{C - A}{C} \times 100.$$

Percent microbial reduction day 12, fifth basin wash compared to the pooled baseline samples from the fifth washes:

$$\text{Fifth wash} = \frac{D - B}{D} \times 100.$$

k. Statistical Analysis

Exploratory data analysis (EDA), including stem-leaf ordering, letter value plots, and box plots, will be generated to assure that the data (R values of the microbial counts) will approximate a normal (Gaussian) distribution [91]. Any outliers will be noted. If the distribution is normal, a series of Student's t tests will be conducted and confidence intervals generated for each of the sample periods using the 0.05 level of significance for type I (α) error.

A pre-post experimental design will be used to evaluate the antimicrobial effectiveness of the extended antimicrobial product use period.

Pre-Product Application			Post-Product Application
$R\ O(1)_{BL1}$	$O(5)_{BL1}$	$O(8)_{BL1}$	$A\ O\,(12)_{T1}$
$R\ O(1)_{BL5}$	$O(5)_{BL5}$	$O(8)_{BL5}$	$A\ O\,(12)_{T5}$

where:

- R = human subjects randomly assigned to the study.
- A = independent variable (test antimicrobial soap).
- $O(1)$ = dependent variable (microbial hand counts), day 1, for baseline (BL) or test (T) for i^{th} wash.

IV. GENERAL-USE HANDWASH EVALUATION

A. EVALUATING ANTIMICROBIAL BODY SOAPS/LOTIONS

There are two general procedures for evaluating these products: (1) the axilla cup scrub method and (2) the full body shower wash.

1. Axilla Cup Scrub Method

For evaluating body soaps and lotions, the most common test in the industry is the axilla cup scrub method. The objective of this evaluation is to determine the ability of the product to keep the microbial populations below a baseline number over periods of both 12 and 24 hours after a single-use application of the antimicrobial product. The underarm is used as the sample site because of the relatively large number of microorganisms that naturally reside there.

Both underarm regions are used. A single product can be applied to both sites or one product at the left side and a second at the right. Only the underarms are washed with the product for this procedure. Generally, the protocol does not require wearing sterile gauze at the evaluation sites to prevent cross-contamination. If the product requires more than a single application to be effective or acquires greater efficacy with prolonged and continuous use, this evaluation may not be applicable.

2. Full Body Shower Wash Evaluation

By the second printing of this book, these products are applied by individuals undergoing surgery, but the FDA has not approved a method for evaluating their efficacy.

A full body shower/bath wash evaluation is a modified version of the full-body presurgical wash evaluation. The axilla and/or the inguinal regions are used for testing in this evaluation. Subjects have baseline samples collected at the test sites on baseline days 1 and 5 [56].

On the first day of the test week, subjects use the product to shower in water at a controlled temperature for a specified length of time. They are immediately sampled by cup scrub at each test site upon completion of the shower. The test sample site is then protected with a sterile, nonocclusive bandage to assure no cross-contamination occurs.

Cup scrub samples are then collected at 12 and 24 hours postshower. Subjects continue to shower using the product for an additional 4 days. On the fifth day of the test, subjects are again sampled immediately after the shower, as well as 12 and 24 hours postshower.

Using this evaluation, an accurate and reliable baseline population can be determined; an immediate postshower measurement is provided in addition to the persistence evaluations of the axilla cup scrub method. Also, the residual properties of the product can be determined by virtue of its repeated use for 5 consecutive days.

B. SAMPLE PROTOCOL: SINGLE APPLICATIONS OF TWO BAR SOAP TEST PRODUCTS

1. Purpose

The objective of this evaluation is to determine the efficacy of single applications of two bar soap test products in reducing the levels of microbial flora residing in the axillary regions of subjects. Test subjects will refrain from using antibacterial/antimicrobial soaps, medicated lotions and creams, underarm deodorants/antiperspirants, and dandruff shampoos throughout the course of the study. Subjects will not bathe or shower within 24 hours of the test product applications or laboratory sampling. Subjects will be provided a kit for their use containing personal hygiene products that do not compromise testing. The study consists of a 15-day washout period during which subjects use only products from the supplied kit and are sampled at the axillae on day 12, using the cup scrub method. On day 16, each test subject will utilize the two test products, one for each of their underarms. At 12 and 24 hours after product application, each subject will have each of their axillae sampled using the cup scrub procedure.

2. Scope

This test method is designed to determine the ability of two antimicrobial soap products to reduce the microbial flora residing in the axillary regions.

3. Test Material

The products to be evaluated are

Test Product 1
 Lot Number: _____
 Expiration Date: _____

Test Product 2
 Lot Number: _____
 Expiration Date: _____

4. Test Methods

a. Subjects

A sufficient number of overtly healthy subjects over the age of 18, but under the age of 70, will be admitted into the study to ensure that 20 subjects complete testing. Subjects will be of mixed sex and age; all will be free from clinically evident dermatoses or injuries to the underarms. All subjects will be provided with study descriptions and will sign Informed Consent Forms prior to participating in the study.

b. Concurrent Treatment

No subject will be admitted into the study who is currently using topical or systemic antimicrobials, or any other medication known to affect the normal microbial flora of the skin.

c. Pretest Period

The 15-day period prior to the product-use portion of the study will be designated the pretest period. During this time, subjects will avoid the use of medicated soaps, lotions, shampoos, deodorants, and so on, as well as skin contact with solvents, acids, and bases. *Deodorants/antiperspirants must not be used in the underarm area by study participants throughout the course of this study.* Subjects will be instructed to use only the personal hygiene products provided to them while participating in the study. Subjects will also avoid using UV tanning beds and bathing in pools or hot tubs containing chlorine or other biocides. Subjects must not shave the axillary sites to be treated for 5 days prior to using the assigned test products. The subjects will not be allowed to bathe or shower within the 24-hour period prior to their sampling times. This regimen will allow for the stabilization of the normal microbial flora of the skin.

Before the initiation of this study, the protocol study description will be given to subjects. Trained laboratory personnel will explain the study to each participant and will be available to answer any questions that may arise. Informed Consent Forms and any other supportive material relevant to the safety of the subjects will be supplied by principal investigators, following their review and approval by an IRB. The primary purpose of the IRB is the protection of the rights and welfare of the subjects involved in the study (reference CFR 21, Parts 50 and 56). This study will begin only after IRB approval of all documents and procedures has been obtained.

Baseline samplings of the axillae will be performed on day 12 of the pretest period utilizing the cup scrub sampling technique.

d. Test Period

On day 16, subjects will be washed with one of the two test antimicrobial soaps at the right axilla first (which product at which site randomly assigned), and then with the second product at the left axilla, following the label instructions provided. T-shirts will be provided for subjects to wear for the duration of testing. The t-shirt must be worn at all times by the subject and must not be washed.

Twelve and 24 hours after the products have been applied, both axillae will be sampled using the cup scrub sampling technique. The 12-hour sample will be taken from the lower portion of the axilla and the 24-hour sample will be taken from the upper portion of the axilla. The two sampling sites must not overlap.

e. Cup Scrub Sampling Technique

The cup scrub sampling technique is performed as follows: Five milliliters (5 mL) of Sterile Stripping Suspending Fluid, consisting of 0.1% Triton X-100 in 0.1 mL/L phosphate-buffered saline solution (pH 7.8), will be added to the cylinder. The skin area inside the cylinder will then be massaged in a circumferential manner with a sterile rubber policeman for 2 minutes. One-milliliter (1 mL) aliquots of this solution are removed and plated in duplicate at appropriate dilutions on soybean-casein digest agar with neutralizers.

The plates will be incubated at $30 \pm 2°C$ for 48 to 72 hours. Those plates providing colony counts between 25 and 250 will be preferentially utilized in this study. If no plates provide counts in the 25 to 250 range, those plates having counts closest to that range will be used in determining the number of viable microorganisms. If 10^0 plates give an average count of zero, the average plate count will be expressed as 1.00. This is done for mathematical reasons. In $\log_{10}$ scale, the $\log_{10}$ of zero is undefined, but the $\log_{10}$ of 1.00 is zero.

f. Methods of Analysis

The plate count data collected from this study will be evaluated using the MiniTab7 (or other) statistical computer software.

The estimated $\log_{10}$ number of viable microorganisms recovered from the axilla samples will be designated the R value. It is the adjusted average $\log_{10}$ colony count measurement for each subject at each sampling time. Each R value will be determined using the following formula:

$$R = \log_{10}\left(\frac{5 \times C_i \times 10^{-D}}{A}\right),$$

where:

A = inside diameter of the cup scrub sampling cylinder.
5 = the number of mL of sterile stripping suspending fluid instilled.

C_i = the arithmetic average colony count of the duplicate plate counts for each subject at a particular dilution level.

D = the dilution factor of the rinsate.

Note: A log_{10} transformation will be performed on the collected data to convert it to a linear scale. A linear scale, more appropriately a log_{10} linear scale, is a basic requirement of the statistical models used in this study.

g. Statistical Analysis

The MiniTab (or other) statistical computer package will be used for all statistical calculations. An EDA, including stem-leaf ordering, letter value plots, and box plots, will be generated to assure that the R values approximate a normal distribution [71]. If this is the case, a two-factor ANOVA test will be conducted and confidence intervals determined for the sample periods using the 0.05 level of significance for type I (α) error. Any outlier values obtained in the analysis will be noted. If a laboratory error is determined to be the cause, these data will not be used in the statistical evaluation; however, they will be included in the raw data.

C. SAMPLE PROTOCOL: FOOD-HANDLER ANTIMICROBIAL HANDWASH EVALUATION

1. Purpose

The objective of this test is to evaluate the antimicrobial efficacy of one handwash product formulation in reducing the level of transient microbial flora (contaminants) on the hands after a single, as well as multiple, handwashes. Fifteen (15) human subjects will be used in this evaluation. The study will consist of a 7-day washout period followed by a 1-day treatment (product-use) period. Enrolled subjects will be instructed to refrain from using antibacterial/antimicrobial soaps, medicated lotions and creams, as well as dandruff shampoos until the study has been completed. Subjects will be provided a personal hygiene kit containing products to be used that do not compromise testing. During the treatment period, subjects' hands will be contaminated with *Escherichia coli*, a generally benign bacterial species distinguishable from the other bacterial flora on the hands through use of differential culture media. Subjects' hands will be contaminated 11 consecutive times and sampled at baseline and three times over the course of 10 contamination/test product wash cycles. The first contamination/sample cycle will be to determine the baseline count using the glove juice sampling method, followed by a handwash with bland (nonantimicrobial) soap. The subject's hands will be contaminated a second time, followed by a handwash with the test product and a glove juice sample. The second contamination and sample cycle will be conducted after the fifth product utilization wash (contamination cycle 6) and the final cycle after the tenth product utilization wash (contamination cycle 11).

2. Scope

This test method is designed to determine the ability of an antimicrobial product formulation to reduce transient microbial flora (contaminants) when used in a handwashing procedure utilizing the transient marker microorganism, *Escherichia coli*.

3. Test Material

The product to be evaluated is

Test Product
    Lot Number: _____
    Manufacture Date: _____

Placebo Product
    Lot Number: _____
    Manufacture Date: _____

4. Test Methods

a. Subjects

A sufficient number of overtly healthy subjects at least the age of 18, but under the age of 70, will be admitted into the study to ensure that 15 subjects complete the study. Subjects will be of mixed sex and age. On the day of testing, all subjects will be free from clinically evident dermatoses or serious injuries to the hands and forearms. All subjects will be thoroughly instructed in the study design and will sign the Informed Consent Forms prior to participating in the study.

b. Concurrent Treatment

No subject will be admitted into the study who is currently using topical or systemic antimicrobials, or any other medication known to affect the normal microbial flora of the skin.

c. Pretest Period

The 7 days prior to the test portion of the study will constitute the pretest period. During this time, subjects will avoid the use of medicated soaps, lotions, deodorants, and shampoos, as well as skin contact with solvents, detergents, acids, and bases. Subjects will be supplied a personal hygiene kit containing a bland soap, shampoo, deodorant, and lotion, as well as a pair of rubber gloves. Subjects must use the contents of this kit exclusively for personal hygiene needs during their participation in the study. They must avoid contact with products on the restricted list. The rubber gloves must be worn when exposure to antimicrobials is unavoidable. Subjects will also avoid using UV tanning beds and bathing in biocidally treated (e.g., chlorinated) pools or hot tubs. This regimen will allow for stabilization of the normal microbial populations residing on the hands.

Before the initiation of this study, the protocol study description will be given to subjects. Informed Consent Forms and any other supportive material relevant to the safety of the subjects will be supplied by the principal investigators, following their review and approval by an IRB. The primary purpose of the IRB is the protection of the rights and welfare of the subjects involved (reference CFR 21, Parts 50 and 56).

A study description and Informed Consent statements will be provided to each subject prior to their beginning the study. Trained laboratory personnel will explain the study to each participant and will be available to answer any questions that may arise.

d. Neutralization

A neutralization assay will be performed to assure that the neutralizers employed in this evaluation adequately inactivate the antimicrobial properties of the test product. If desired, a wash with a surgical scrub product may be added as well.

e. Test Period

A period of 7 days will constitute the test week. Each subject will be employed 1 day of that week for about a 2-hour test period. A practice wash using the nonantimicrobial bland soap and the application procedure prescribed for the test product will precede the actual test portion of this study. The practice wash ensures that the subject understands the wash procedure.

After the practice wash has been completed, a 5.0 mL aliquot of a suspension containing approximately 1.0×10^8 CFU/mL of *Escherichia coli* (ATCC #11229) will be transferred into each subject's cupped hands. The inoculum will then be distributed evenly over both hands and up to the wrist, via gentle continuous massage for 45 seconds. After a 2-minute air-dry, the glove juice sampling procedure will be performed. This first inoculation cycle will provide the baseline inoculation levels. It will be followed by a 30-second nonmedicated soap handwash. The temperature of the water used for this and all subsequent wash cycles will be controlled at $40 \pm 2°C$.

The microbial inoculum will again be distributed over both hands up to the wrist, via gentle continuous massage for 45 seconds. After a 2-minute air-dry, the 15 subjects will wash with the test product according to the prescribed use directions. This will be followed by the glove juice sampling procedure.

This inoculation/wash procedure will be repeated a total of 11 consecutive times with a minimum of 5 and a maximum of 15 minutes between applications. Samplings of the hands for residual *Escherichia coli* will be performed after inoculation/product wash cycles 1, 5, and 10 using the glove juice sampling procedure.

f. Glove Juice Sampling Procedure

The glove juice sampling procedure will be performed as follows: After the prescribed wash and rinse, the hands will be lightly dried with disposable paper towels and powder-free sterile gloves will be donned. At the designated sampling times, 75.0 mL of Sterile Stripping Suspending Fluid without product neutralizers will be instilled into the glove. The wrist will be secured, and an attendant will massage the hand through the glove in a standardized manner for 60 seconds. Aliquots of the glove juice (dilution 10^0) will be removed and serially diluted in 0.1% *M* phosphate-buffered solution with appropriate neutralizers.

Triplicate spread plates will be prepared from each of these dilutions, using MacConkey agar with neutralizers. The *Escherichia coli* will produce purple colonies with a greenish metallic sheen on this differential medium, distinguishing it from the yellow colonies of the normal skin flora. The plates will be incubated at $30 \pm 2°C$ for approximately 48 hours. Those plates providing colony counts of purple colonies between 25 and 250 will be preferentially utilized in this study. If no plates provide counts in the 25 to 250 range, the plates closest to that range will be counted and used in determining the number of viable microorganisms. If 10^0 plates give an average count of zero, the average plate count will be expressed as 1.00. This is done for mathematical reasons. In $\log_{10}$ scale, the $\log_{10}$ of zero is undefined, but the $\log_{10}$ of 1.00 is zero. The number of viable bacteria recovered is obtainable from the formula $75 \times$ Dilution Factor $\times$ Mean Plate Count for the three plates.

Following the final product wash/sample cycle, the subjects will be required to wash with 70% isopropyl alcohol for 1 minute, air-dry the hands, and rinse under tap water in order to remove any remaining *Escherichia coli* from the hands.

g.   Methods of Analysis

The plate count data collected from this study will be entered and evaluated using MiniTab (or other) statistical computer software.

The estimated $\log_{10}$ number of viable microorganisms recovered from each hand will be designated the R value. It is the adjusted average $\log_{10}$ colony count measurement for each subject at each sampling time. Each R value will be determined using the following formula:

$$R = \log_{10}\left[75 \times C_i \times 10^{-D} \right],$$

where:

   75 = the amount of stripping solution instilled into each glove.
   C_i = the arithmetic average colony count of the triplicate plate counts for each subject at a particular dilution level.
   D = the dilution factor.

Note: The $\log_{10}$ transformation is performed on these data to convert them to a linear scale. A linear scale, more appropriately a $\log_{10}$ linear scale, is a requirement of the statistical models to be used.

h.   Statistical Analysis

A pre-/post-experimental design will be utilized to evaluate and compare the antimicrobial effectiveness of the test product:

Pre-Product Application	Post-Product Application
$R\ A\ O_{BL}$	$O_1\ O_5\ O_{10}$

where:

   R = subjects randomly assigned to the study.
   A = independent variable: liquid test formulation.
   o_i = dependent variable: microbial counts at baseline (BL) and after product-use washes 1, 5, and 10.

EDA will be performed on the data. Stem-leaf ordering, letter value plots, and box plots will be generated to assure the data collected approximate normal distribution [91]. If this is the case, a two-factor ANOVA test will be conducted using the 0.05 level of significance for type I (α) error. Any outlier values will be noted. If a laboratory error is determined to be the cause, those data will not be used in the statistical evaluation, but they will be included among the raw data.

V. MEDICAL/HEALTHCARE INDUSTRY ANTIMICROBIALS

In this area, the most frequently used topical antimicrobial studies are the healthcare personnel handwash, the surgical scrub, and the preoperative skin-prepping evaluations. These evaluations have received a great deal of input from the FDA and are presented in detail in the Tentative Final Monograph (CFR 21, Parts 333 and 369, Part III; 17 June 94) [112]. Examples of the basic working protocols follow.

A. SAMPLE PROTOCOL: HEALTHCARE PERSONNEL HANDWASH

1. Purpose
The purpose of this study is to examine the antimicrobial efficacy of one healthcare personnel handwash test product, one vehicle (test product without active ingredient), and one control product.

2. Scope
The antimicrobial effectiveness of one test healthcare personnel handwash formulation, one vehicle, and one control product will be determined utilizing 30 human subjects per formulation (a total of 90 subjects) over the course of 10 handwashes, with microbial samples taken after washes 1, 3, 7, and 10.

3. Test Material
The products to be evaluated are

> Test Product
> > Lot Number: _____
> > Manufacture Date: _____

> Vehicle Product
> > Lot Number: _____
> > Manufacture Date: _____

> Positive Control Product
> > Lot Number: _____
> > Manufacture Date: _____

4. Test Methods

a. Subjects

A sufficient number of overtly healthy subjects over the age of 18, but under the age of 70, will be admitted into the study to ensure that 90 subjects complete the study

(30 subjects per product configuration). Subjects will be of mixed sex and age and free from clinically evident dermatoses, open wounds, injuries to the hands or forearms, hangnails, or other disorders that may compromise the subject. All subjects will be familiarized with the study protocol and will sign Informed Consent Forms prior to participating in the study.

b. Concurrent Treatment

No subject will be admitted into the study who is currently using topical or systemic antimicrobials, or any other medication known to affect the normal microbial flora of the skin.

c. Pretest Period

The 7 days prior to the test portion of the study will comprise the pretest period. During this time, subjects will avoid the use of medicated soaps, lotions, deodorants, and shampoos as well as skin contact with solvents, detergents, acids, and bases. Subjects will be supplied a personal hygiene kit containing nonmedicated soap, shampoo, deodorant, lotion, and latex gloves to be worn when contact with restricted materials cannot be avoided. Subjects must use the contents of this kit exclusively during their participation in the study. They must avoid contact with products on the list of restricted materials. Subjects will also avoid using UV tanning beds and bathing in chlorinated pools or hot tubs. This regimen will allow for stabilization of the normal microbial populations residing on the hands.

Before the initiation of this study, the protocol study description will be given to subjects. Trained laboratory personnel will explain the study to each participant and will be available to answer any questions that may arise. Informed Consent Forms and any other supportive material relevant to the safety of the subjects will be supplied by principal investigators, following their review and approval by an IRB. The primary purpose of the IRB is the protection of the rights and welfare of the subjects involved in the study (reference CFR 21, Part 56). This study will begin only after IRB approval of all materials and procedures has been obtained.

d. Neutralization

A neutralization assay will be performed to assure that the neutralizers employed effectively neutralize the antimicrobial activity of the product in this evaluation.

e. Experimental Period

The test period will comprise a 14-day period. Each subject will be employed only 1 day of that period for a 4- to 5-hour period. A 30-second practice wash will be performed using a nonantimicrobial (bland) soap and the same application procedure intended for that subject's test product. The practice wash will ensure that the subject understands the wash procedure. The temperature of the water used for this and all subsequent wash cycles will be controlled at $40 \pm 2°C$.

On the designated test day, a 5.0 mL aliquot of a suspension containing at least 1.0×10^8 CFU/mL *Serratia marcescens* (ATCC# 14756, red-pigmented strain) will be transferred into each subject's cupped hands. The inoculum will then be distributed evenly over both hands and not reaching above the wrists, via gentle continuous massage for 45 seconds. After a 2-minute air-dry, the glove juice sampling procedure will be performed. This inoculation/sampling cycle will provide the baseline inoculation levels. It will be followed with a 30-second bland soap handwash.

The microbial inoculum will again be distributed evenly over both hands, and not reaching above the wrists, via gentle continuous massage for 45 seconds. After a 2-minute air-dry, the subjects will wash with their assigned product configuration as specified by the sponsor-supplied use directions. This test wash will be followed by the glove juice sampling procedure.

This inoculation/product wash procedure will be repeated 10 consecutive times, with a minimum of 5 minutes and a maximum of 15 minutes between applications. Samplings of the hands for remaining *S. marcescens* will be performed after inoculation/product wash cycles 1, 3, 7, and 10, using the glove juice sampling procedure.

f. Glove Juice Sampling Procedure

The glove juice sampling procedure will be performed as follows: Following the prescribed wash and rinse, excess water will be shaken from the hands and powder-free, loose-fitting sterile latex gloves will be put on. At the designated sampling time, 75 mL of Sterile Stripping Suspending Fluid without product neutralizers will be instilled into the glove. The wrist will be secured and an attendant will massage the hand through the glove in a uniform manner for 60 seconds. Aliquots of the glove juice (dilution 10^0) will be removed and serially diluted in Butterfield's phosphate-buffered diluent solution with appropriate product neutralizers.

Spread plates will be prepared in triplicate from each of these dilutions using soybean-casein digest agar and incubated at $25 \pm 2°C$ for approximately 48 hours. Those plates providing colony counts between 25 and 250 will be preferentially utilized in this study. If no plates provide counts in the 25 to 250 range, those plates with counts closest to that range will be used in determining the number of viable microorganisms. If 10^0 plates give an average count of zero, the average plate count will be expressed as 1.00. This is done for mathematical reasons. In $\log_{10}$ scale, the $\log_{10}$ of zero is undefined, but the $\log_{10}$ of 1.00 is zero. The number of viable microorganisms recovered is obtainable from the formula

$$75 \times \text{Dilution Factor} \times \text{Mean Plate Count}$$

for the three plates.

Following the final product application cycle (application 10), the subjects will be required to wash with 70% isopropyl alcohol for 1 minute, air-dry their hands, and rinse under tap water in order to remove any remaining *Serratia marcescens* from the hands. If desired, a wash with a surgical scrub product may be added as well.

g. Methods of Analysis

The plate-count data collected from this study will be evaluated using MiniTab (or other) statistical computer software. The estimated $\log_{10}$ number of viable microorganisms recovered from each hand will be designated the R value. It is the adjusted average $\log_{10}$ colony count measurement for each subject at each sampling time. Each R value will be determined using the following formula:

$$R = \log_{10}\left[75 \times C_i \times 10^{-D}\right],$$

where:

    75 = the amount of stripping solution instilled into each glove.
    C_i = the arithmetic average colony count of the three plate counts for each subject at a particular dilution level.
    D = the dilution factor.

Note: A $\log_{10}$ transformation will be performed on these data to convert them to a linear scale. A linear scale, more appropriately a $\log_{10}$ linear scale, is a requirement of the statistical models to be used.

h.   Statistical Analysis   A pre-/post-experimental design will be utilized to evaluate and compare the antimicrobial effectiveness of the test product.

Pre-Product Application	Post-Product Application
R $A(1)$ $O(1)_{BL}$	$O(1)_1$ $O(1)_3$ $O(1)_7$ $O(1)_{10}$
R $A(2)$ $O(2)_{BL}$	$O(2)_1$ $O(2)_3$ $O(2)_7$ $O(2)_{10}$
R $A(3)$ $O(3)_{BL}$	$O(3)_1$ $O(3)_3$ $O(3)_7$ $O(3)_{10}$

where:

    R = subjects randomly assigned to one of three products in this study.
    $A(i)$ = independent variables: the product configuration "i" assigned to each subject, where $i = 1$ for test product, 2 for vehicle, and 3 for control product.
    $O(1)_i$ = dependent variables: microbial counts at baseline (BL) and after the ith product use (washes 1, 3, 7, and 10).

Because only washes 1 and 10 have any merit with the FDA (i.e., a 2.00 $\log_{10}$ reduction after wash 1 and a 3.00 $\log_{10}$ reduction after wash 10), use only these washes in the statistical analysis. Do not use washes 3 or 7.

Prior to performing a statistical analysis, EDA will be performed on the data. Stem-leaf ordering, letter value displays, and box plots will be generated to assure the data collected approximate the normal (Gaussian) distribution. A two-factor ANOVA test will be conducted using the 0.05 level of significance for type I (α) error [30]. Any outlier values will be noted. If a laboratory error is determined to be the cause, those data will not be used in the statistical evaluation, but they will be included among the raw data.

B. SAMPLE PROTOCOL: SURGICAL SCRUBS

1. Purpose
The purpose of this study is to evaluate and compare the antimicrobial efficacy of three surgical scrub product configurations.

2. Scope
This study will evaluate and compare the antimicrobial efficacy of three surgical scrub product configurations. Antimicrobial efficacy will be measured in terms of three parameters—the immediate, the persistent, and the residual antimicrobial effects. At least 18 human subjects will be employed for each of the three product configurations for a total of 54 subjects.

3. Test Materials
Products to be used in this study are

Test Product
    Lot Number: _____
    Expiration Date: _____

Vehicle Product
    Lot Number: _____
    Expiration Date: _____

Reference Product
    Lot Number: _____
    Expiration Date: _____

4. Test Methods

a. Subjects

A sufficient number of overtly healthy subjects over the age of 18, but under the age of 70, will be admitted into the study to ensure that 54 subjects complete the study (18 subjects per product configuration). Subjects will be of mixed sex and age and free of clinically evident dermatoses or injuries to the hands or forearms. All subjects will be familiarized with the study protocol and will sign Informed Consent Forms prior to participating in the study.

b. Concurrent Treatment

No subject will be admitted into the study who is currently using topical or systemic antimicrobials, or any other medication known to affect the normal microbial flora of the skin.

c. Pretest Period

The 14 days prior to the baseline portion of the study will constitute the pretest period. During this time, subjects will avoid the use of medicated soaps, lotions, deodorants, and shampoos, as well as skin contact with solvents, detergents, acids, and bases. Throughout the course of the study, subjects are to use only the non-antimicrobial personal products that are supplied to them. This regimen will allow for stabilization of the normal microbial populations residing on the hands.

d. Neutralization

A neutralization study will be performed prior to beginning actual testing in order to assure that the neutralizer(s) used effectively neutralize the antimicrobial properties of the products.

e. Baseline Period

Baseline measurements will be conducted over a week's time (baseline week) on days 1, 3, and 5, using a nonantimicrobial (bland) soap. The baseline count of the resident microbial populations will be used to evaluate eligibility for the study, as well as establish baseline values for each subject. Only those subjects with baseline counts of at least 1.5×10^5 organisms per hand will be selected to continue the study.

The baseline determinations will utilize the same sampling and recovery techniques used for the experimental period. Both hands will be sampled in baseline determinations. Subjects will not wash their hands for at least 2 hours prior to baseline samplings. Subjects will clean under fingernails with a nail stick and clip fingernails to a free-edge of less than or equal to 2 mm. All jewelry must be removed from the hands and arms.

Baseline determinations will be conducted as follows: The hands and the forearms up to two thirds the distance from wrist to elbow will be washed in at $40 \pm 2°C$ water for 30 seconds using a bland soap. Samples will then be taken immediately (within 1 minute of scrubbing) using the glove juice sampling procedure.

f. Experimental Period

Subjects accepted into the study will be randomly assigned to one of the three product groups and be sampled as outlined in Tables 10.1 and 10.2. The test products will be used according to the directions supplied by the sponsor. The water temperature used for the surgical hand scrub will be regulated at $40° \pm 2°C$. The subjects will wash with the assigned products once on test days 1 and 5 and three times on test days 2, 3, and 4. The hands will be sampled after the first scrub procedure on days 1,

---

TABLE 10.1
Surgical Scrub Schema

Test Day	1	2	3	4	5
Number of Scrubs per Day	1	3	3	3	1

Note: Left and right hands will be randomly sampled at times indicated in the sampling schedule using the glove juice sampling procedure.

TABLE 10.2

Sampling Schedule: Number of Hands Sampled at Indicated Time after Treatment

		Hour		
	Day	0	3	6
Baseline Week: Baseline	1, 3, and 5	108	0	0
Test Week: Test Product	1, 2, and 5	12	12	12
2% Chlorhexidine Gluconate				
Test Week: Vehicle Product	1, 2, and 5	12	12	12
Test Product w/o the 2% Chlorhexidine Gluconate				
Test Week: Reference Product	1, 2, and 5	12	12	12
4% Chlorhexidine Gluconate				

2, and 5. On each of the sampling days, the subjects' hands will be sampled at time zero (immediately after the scrub), 3 hours post-scrub, and 6 hours post-scrub.

Note: The surgical scrub schema and the sampling schedule allow for the assessment of the immediate, persistent, and residual effectiveness of the test products in reducing the normal microbial flora of the hands.

g.   Glove-Juice Sampling Procedure

Following the prescribed wash and rinse, excess water will be shaken from the hands and loose-fitting, powder-free sterile gloves donned. At the designated sampling time, 75 mL of Sterile Stripping Suspending Fluid with neutralizers (pH 7.8 to 7.9) will be instilled into the glove. The wrist will be secured, and an attendant will massage the hand through the glove in a standardized manner for 60 seconds, paying particular attention to the area under the nails. Aliquots of the glove juice (10^0) will be removed and serially diluted in Butterfield's phosphate-buffered diluent with neutralizers. Pour plates of soybean-casein digest agar with neutralizers will be prepared in duplicate with 1.0 mL aliquots of these dilutions. The solidified plates will be incubated at $30 \pm 2°C$ for approximately 48 hours. Those plates providing colony counts between 25 and 250 will be preferentially utilized in this study. If no plates provide counts in the 25 to 250 range, the plates having counts closest to that range will be used in determining the number of viable microorganisms. If 10^0 plates give an average count of zero, the population will be expressed as <10 rather than <25. The number of viable bacteria recovered is obtainable from the formula

$$75 \times \text{dilution factor} \times \text{mean plate count.}$$

h.   Statistical Analysis

A two-factor ANOVA statistical model will be utilized to determine and compare the antimicrobial effectiveness of the test products. The 0.05 level of significance should be used in this study.

C. SAMPLE PROTOCOL: PREOPERATIVE SKIN-PREPPING SOLUTIONS

1. Purpose

The purpose of this study is to evaluate and compare the antimicrobial effectiveness of one preoperative skin-prepping formulation and one control product. There may be more products, however. If an alcohol and chlorhexidine gluconate (CHG) product is used, then the configuration would be: (1) test product, (2) alcohol, (3) CHG, and (4) control product.

2. Scope

A minimum of 30 human subjects will be employed, 15 subjects per each of the products, utilizing bilateral product applications. Because the lower bounds of the 95% confidence interval must be ≥ 3.00 $\log_{10}$ at the inguinal site and ≥ 2.00 $\log_{10}$ at the abdominal site, be certain to allow for a greater number of subjects. There should be a pilot study performed to see where the data lie. The two separate anatomical sites, the inguinae and the abdomen, will be used in evaluating the product's immediate antimicrobial effects at 10 minutes and their persistent antimicrobial effects at 30 minutes and 6 hours post-skin-prepping. These configurations are often changed to 30 seconds, 10 minutes, and 6 hours.

3. Test Material

The following test materials will be used in this evaluation:

    Test Products
        Product 1: (test formulation)
            Lot Number: _____
            Expiration Date: _____
            Manufacture Date: _____

        Product 2: (reference formulation)
            Lot Number: _____
            Expiration Date: _____
            Manufacture Date: _____

a. Subjects

A sufficient number of overtly healthy subjects of at least the age of 18, but under the age of 70, will be admitted in the study to ensure that at least 30 subjects (15 per product configuration) complete the evaluation. Insofar as possible, subjects will be of mixed age, sex, and race, and all will be free of dermatoses or injuries to the skin areas being sampled.

b. Concurrent Treatment

No subject will be admitted into this study who is currently using a known topical or systemic antimicrobial, or any other medication known to affect the normal microbial flora of the skin.

c. Pretest Period

The 7-day period prior to the product-use (test) portion of the study will be designated the pretest period. Subjects will be provided a personal hygiene kit that must be used for all personal care throughout the course of the study. Subjects will use only the assigned personal care soaps, lotions, shampoos, deodorants, and so on, as well as avoid skin contact with solvents, acids, and bases. In addition, bathing in pools or hot tubs containing chlorine or other biocides will be prohibited. Subjects must not shave the anatomical sites proposed for treatment within the 5 days prior to being prepped. The subjects will not be allowed to bathe or shower within the 24-hour period prior to their sampling times. This regimen will allow for the stabilization of the normal microbial flora of the skin.

Before the initiation of this study, the protocol, study description, the subject information sheets, Informed Consent Forms, and any other materials relevant to the subjects' safety will be reviewed and approved by an IRB. The primary purpose of the IRB is to protect the rights and welfare of the subjects involved.

A description of this study and an informed consent statement (patient information) will be provided to subjects prior to their beginning the study. Trained laboratory personnel will explain the study to each participant and will be available to answer any questions that may arise.

d. Neutralization

A neutralization study will be performed prior to beginning actual testing in order to assure that the neutralizer(s) used effectively neutralize the antimicrobial properties of the products.

e. Baseline Week

Baseline sampling bilaterally in the inguinal area and on the abdomen in the vicinity of and on both sides of the umbilicus will be performed on 2 separate days of the baseline week, utilizing the cylinder sampling technique. The first baseline samples will be collected on day 1 of the baseline week and the second on day 4. In order for a particular subject to be accepted into the study, baseline colony counts at the inguinae must be at least 1.0×10^5 microbial organisms per cm^2. Baseline colony counts on the abdomen must be at least $1.0 \times 10^3/cm^2$.

f. Experimental Period

Subjects accepted into the study will be randomly assigned one of the two products. Prior to testing, the subjects will be questioned regarding adherence to the protocol and physically examined to ensure no evidence of injury or dermatosis is present at the sampling sites.

Subjects will then be sampled for baseline counts bilaterally at both the inguinae and the abdomen in the vicinity of the umbilicus. The inguinal and abdominal sites will be sampled two times for baseline. Subjects will then be prepped with the assigned product at the inguinae and the abdomen near the umbilicus following the label instructions provided.

A sample of each prepped area will be taken at 10 minutes post-prepping using the cylinder sampling technique. After this sample, a gauze and fenestration bandage (Smith & Nephews, Opsite I.V.) will be placed over the prepped areas to prevent any microbial contamination. Additional samples will be taken at 30 minutes and 6 hours post-skin-prepping.

g. Cylinder Sampling Technique

The cylinder sampling technique will be performed as follows: At the designated sampling times, a sterile cylinder will be held firmly onto the skin surface of the test site to be sampled, and 2.5 mL of sterile 0.1 M phosphate-buffered stripping solution with 0.1% Triton X-100 and product neutralizers will be instilled into the cylinder; the skin area inside the cylinder will be massaged in a circumferential manner for 1 minute with a sterile rubber policeman. The entire procedure will be repeated without lifting the cylinder to obtain a second aliquot, and the two aliquots of 2.5 mL each will then be pooled.

One-milliliter (1 mL) aliquots of the microorganism suspension (10^0 dilution) will be removed, plated, and/or serially diluted (as appropriate) with Butterfield's phosphate-buffered solution. Pour plates will be prepared in triplicate with 1 mL from each of these dilutions using soybean-casein digest agar and the plates will be incubated at $30 \pm 2°C$ for 72 ± 2 hours. Those dilutions yielding 25 to 250 colonies per plate will be preferentially counted. In cases where no plates provide counts in the 25 to 250 range, those plates with the highest number of countable colonies closest to that range or those from the lowest dilution will be used.

h. Data Handling and Statistics

The estimated $\log_{10}$ number of viable microorganisms per cm^2 recovered from each sample site will be designated the R value. To convert the volumetric quantities collected during sampling into the number of colony-forming units per square centimeter (cm^2), the following formula will be employed:

$$R = \log_{10}\left[\frac{F\left(\dfrac{C_i}{n}\right)10^{-D}}{A}\right],$$

where:

R = the average colony count in $\log_{10}$ scale per cm^2 of sampling surface.

F = total number of mL of stripping fluid added to the sampling cylinder. In this study, F is equal to 5 mL.

$\left(\dfrac{C_i}{n}\right)$ = average of the triplicate plate counts used.

D = dilution factor of the plate counts.

A = inside area of the cylinder in cm^2.

Note: A $\log_{10}$ transformation will be performed on the collected data to convert them to a linear scale. A linear scale, more appropriately a $\log_{10}$ linear scale, is a basic requirement of the statistical models used in this study.

i. Statistical Analysis

The MiniTab (or other) statistical computer package will be used for all statistical calculations. An EDA, including stem-leaf ordering, letter value plots, and box plots, will be generated to assure that the R values approximate a normal (Gaussian) distribution [91]. A blocked, two-factor ANOVA test will be conducted and confidence intervals determined for the sample periods using the 0.05 level of significance for type I (α) error.

Any outlier values obtained in the analysis will be noted. If laboratory error is determined to be the cause, these data will not be used in the statistical evaluation; however, they will be included in the raw data.

A pre-post experimental design will be utilized to evaluate and compare the antimicrobial effectiveness of the products.

Pre-Product Application*	Post-Product Application**
$R\ A\ \ O_{A,Bl,1}\ \ O_{A,BL,2}$	$O_{A,T,10min}\ \ O_{A,T,30min}\ \ O_{A,T,6hr}$
$R\ C\ \ O_{C,BL,1}\ \ O_{C,BL,2}$	$O_{C,T,10min}\ \ O_{C,T,30min}\ \ O_{C,T,6hr}$

where:

R = Subjects randomly assigned to each of the products (2, in this case). A random number series will be generated for this purpose.

A = Independent variable: In this study, A represents the test product.

C = Independent variable: In this study, C represents the reference product.

O_{xij} = Dependent variable: For test or control product (A or C), microbial counts per cm^2 at the ith period and the jth time.

x = Test (T) or Control (C) product.

i = Baseline (BL) or Test (T_P) period.

j = Period 1 or 2 for the baseline phase and times of 10 minutes, 30 minutes, or 6 hours for test phase.

D. SAMPLE PROTOCOL: INDWELLING CATHETER/PREINJECTION PREPARATION

1. Purpose

The purpose of this study is to evaluate and compare the antimicrobial effectiveness of one antimicrobial solution in two different application packages, the vehicle in two different application packages, and one reference product.

2. Scope

This evaluation will measure the antimicrobial effectiveness of one test product presented in two different application packages. One application package is used for indwelling catheter site preparations and the other is used for preinjection site preparations. The vehicle, presented in packages identical to those of the test product, and a reference product are to be used as controls for both the indwelling catheter and the preinjection preparation. A minimum of 45 human subjects (a minimum of 30 sites

---

* There may be four products evaluated, depending on the test product.

** Times can be changed to 30 seconds, 10 minutes, and 6 hours.

per formulation) will be employed, and product applications will be bilateral. The immediate and persistent antimicrobial efficacy of the formulations will be evaluated at three separate anatomical sites over a 24-hour period to support the claim of an indwelling catheter preparation. For the claim of a preinjection preparation, only the immediate antimicrobial efficacy will be evaluated at two separate anatomical sites.

3. Test Material

The products to be used in this evaluation are

    Test Product Applicator 1
        Lot Number: _____
        Expiration Date: _____

    Test Product Applicator 2
        Lot Number: _____
        Expiration Date: _____

    Product Vehicle Applicator 1
        Lot Number: _____
        Expiration Date: _____

    Product Vehicle Applicator 2
        Lot Number: _____
        Expiration Date: _____

    Reference Product
        Lot Number: _____
        Expiration Date: _____

4. Test Sites

The test product in application package 1 will be utilized only at the (a) deltoid and (b) gluteus maximus sites. The test product in application package 2 will be utilized only at the (a) subclavian, (b) femoral (inguinal), and (c) median cubital sites. The vehicle and reference product will be used at all anatomical sites.

Each subject will be randomly assigned one product configuration for bilateral product application. There are three possible product configurations. Both test product applications (test product application 1 utilized at the deltoid and gluteus maximus sites and test product application 2 utilized at the subclavian, femoral, and median cubital sites) and the vehicle configurations; both test product application and the reference; and the reference and vehicle product application.

5. Test Methods

a. *Institutional Review Board*

Prior to the initiation of this study, an IRB will be assembled to review and approve the protocol, including the study description, Informed Consent Forms, and any other supportive material relevant to the safety of the human subjects involved in the study (reference CFR 21, Parts 50 and 56). This study will begin only after IRB approval has been obtained for all facets.

b. Subjects

A sufficient number of overtly healthy subjects of at least the age of 18, but under the age of 70, will be admitted to the study to ensure that at least 45 subjects utilizing bilateral product application complete the evaluation. Insofar as possible, subjects will be of mixed age, sex, and race, and all will be free of dermatoses or injuries to the sites being sampled.

Because both test product applicators will be used at separate sites on one side of each subject (test product applicator 1 at the deltoid and gluteus maximus; test product applicator 2 at the subclavian, femoral, and median cubital regions), each subject will be assigned one of three possible product configurations: (1) test products and vehicle; (2) test products and reference; and (3) reference and vehicle. Hence, each of the 45 subjects will receive one of these test configurations.

Trained laboratory personnel will review the study description with each subject. Upon completion of their review, each subject will be given a copy of the study description and will sign and date the Informed Consent Form and complete the Subject Confidential Information and Acceptance Criteria. Trained laboratory personnel will assign a subject number to each subject and complete the Informed Consent for each subject.

c. Concurrent Treatment

No subject will be admitted into this study who is currently using a known topical or systemic antimicrobial, or any other substance known to affect the normal microbial flora of the skin.

d. Pretest Period

The 14-day period prior to the product-use portion of the study will be designated the pretest period. During this time, subjects will use only the personal hygiene soaps, lotions, shampoos, deodorants, and so on that are provided and will avoid skin contact with solvents, acids, and bases. In addition, subjects will avoid use of tanning booths and bathing in pools or hot tubs containing chlorine or other biocides. The subjects will not be allowed to bathe or shower during the 24-hour period prior to being sampled. This regimen will allow for the stabilization of the normal microbial flora of the skin.

e. Neutralization

Prior to initiation of this study, the adequacy of the neutralizers will be confirmed in accordance with *Standard Practices for Evaluating Inactivators of Antimicrobial Agents* E1054-08 (2013) (ASTM E1054-91) (97).

f. Baseline Week

The week following the 14-day pretest period will constitute the baseline week. Subjects will not bathe or shower within 24 hours of being sampled. All subjects will be sampled on days 1 and 4 of that week at the five selected anatomical sites per bilateral configuration. There will be a minimum of 72 hours between the time the baseline period ends and the experimental period begins. Based upon adequate baseline microbial counts from these two baseline assessments, subjects will be selected to complete this study. Baseline acceptance criteria for deltoid, gluteus maximus, median cubital, and subclavian areas will be at least 5.0×10^2 CFU/cm². Acceptable counts for the femoral (inguinal) region will be at least 1.0×10^5 CFU/cm².

All sample sites requiring clipping will be clipped at least 48 hours prior to test day 1. Clipping will be completed to assure that the bandaging material (used during the experimental test period) remains securely affixed to the test site for the 24-hour testing period. Prior to being sampled, the subjects will be questioned regarding their adherence to protocol restrictions. Also, sampling sites will be examined visually to ensure no evidence of injury or dermatosis is present.

A third and final baseline sample will be collected at each test site prior to prepping/testing on test day 1. Sampling will be performed near the (1) subclavian vein, (2) the median cubital vein of the forearm, (3) the femoral vein in the inguinal area, (4) the deltoid region of the arm, and (5) the gluteus maximus (buttocks) using the cylinder sampling technique, depending where the product is used.

g.   Test Period

Subjects will be questioned regarding their adherence to protocol restriction requirements and examined visually to ensure no evidence of injury or dermatosis is present at the anatomical sampling sites. Subjects will refrain from showering or bathing during the 24-hour period prior to being sampled.

Subjects will not shower or bathe while they are in the 24-hour experimental period. Subjects may take sponge baths, but are to avoid washing the bandaged sampling sites.

h.   Indwelling Catheter Preparation

On day 1 of the test week, the regions near the subclavian vein, the femoral vein, and the median cubital vein will be sampled for the final (third) baseline measurement and then prepped according to the sponsor's instructions using the assigned product. Samples will be taken immediately post-prepping using the cylinder sampling technique. A sterile bandage (Smith and Nephew bandage material) will be placed over the prepped area to help prevent microbial contamination. Additional samples will be taken 24 hours post-prepping using the cylinder sampling technique.

i.   Preinjection Preparation

On day 1 of the test week, the deltoid and gluteus maximus sites will be sampled for the final (third) baseline measurement and then prepped using the assigned product according to the sponsor's instructions. Samples will be taken immediately post-prepping using the cylinder sampling technique.

j.   Cylinder Sampling Technique

The cylinder sampling technique will be performed as follows: At the designated sampling times, a sterile cylinder will be held firmly on the skin surface of the test site to be sampled, and 2.5 mL of sterile 0.1 M phosphate-buffered stripping solution with 0.1% Triton X-100 and product neutralizers will be instilled into the cylinder. The skin area inside the cylinder will be massaged in a circumferential manner for 1 minute with a sterile rubber policeman. The entire procedure will be repeated to obtain a second aliquot, and the two aliquots of 2.5 mL each will be pooled.

One-milliliter (1 mL) aliquots of the microorganism suspension (10^0 dilution) will be removed, plated, and/or serially diluted (as appropriate) with Butterfield's phosphate-buffered solution. Triplicate pour plates will be prepared with 1 mL from each of these dilutions using soybean-casein digest agar.

The inoculated plates will be incubated at $30 \pm 2°C$ for 72 ± 2 hours. Those dilutions yielding 25 to 250 colonies per plate will be preferentially counted. In cases where no plates provide counts in the 25 to 250 range, those plates with the highest number of countable colonies closest to that range or the lowest dilution will be used.

k. Test Duration

Testing will continue over a 1- to 2-month period with three groups of 15 people staggered over that time.

l. Data Handling and Statistics

The plate-count data will be recorded on raw data sheets. The raw plate-count data will be transformed into $\log_{10}$ values to linearize them, a requirement of the statistical models used. The statistical level of significance will be set at $\alpha = 0.05$ and all confidence intervals will be reported at the 95% confidence level. A MiniTab® (or other) computer statistical package will be used for all statistical calculations.

In order to convert the volumetric measurements into the number of colony-forming units per square centimeter (cm^2), the following formula will be employed:

$$R = \log_{10} \left[\frac{F\left(\dfrac{C_i}{n}\right) 10^{-D}}{A} \right],$$

where:

R = the average colony count in $\log_{10}$ scale per cm^2 of sampling surface.

F = total number of mL of stripping fluid added to the sampling cylinder. For this study, F is equal to five (5) mL.

$\left(\dfrac{C_i}{n}\right)$ = average of the triplicate plate counts used for each sample collected.

D = dilution factor of the plate counts.

A = inside area of the cylinder in cm^2.

A pre/post experimental design will be used to evaluate the antimicrobial properties. For the preinjection design, see Figure 10.1. Figure 10.2 presents the indwelling catheter preparation design. A biostatistical evaluation will be performed on the

	Pre-Product Application	Post-Product Application
$R\ T_1$	$O_{BL1}\ O_{BL2}\ O_{BL3}$	O_{T1}
$R\ V$	$O_{BL1}\ O_{BL2}\ O_{BL3}$	O_{T1}
$R\ C$	$O_{BL1}\ O_{BL2}\ O_{BL3}$	O_{T1}

where:

R = human subjects randomly assigned to products.
T_1 = independent variable (test product 1).
V = independent variable (vehicle product).
C = independent variable (reference product).
$O_{i,j}$ = dependent variable (microbial counts/cm^2).
  i = BL if baseline, T if test.
  j = 1, 2, 3 for baseline samples or 1 if test sample (immediate).

FIGURE 10.1   Preinjection design evaluated at the deltoid and gluteus maximus sites.

	Pre-Product Application	Post-Product Application
$R\ T_2$	$O_{BL1}\ O_{BL2}\ O_{BL3}$	$O_{T1}\ O_{T2}\ O_{T3}$
$R\ V$	$O_{BL1}\ O_{BL2}\ O_{BL3}$	$O_{T1}\ O_{T2}\ O_{T3}$
$R\ C$	$O_{BL1}\ O_{BL2}\ O_{BL3}$	$O_{T1}\ O_{T2}\ O_{T3}$

where:

R = human subjects randomly assigned to products.
T_2 = independent variable (test product 2).
V = independent variable (vehicle product).
C = independent variable (reference product).
$O_{i,j}$ = dependent variable (microbial counts/cm^2).
  i = BL if baseline, T if test.
  j = 1, 2, 3 for baseline samples.

  or

  j = 1, 2, 3 for test samples:
    1 if immediate
    2 if 24 hours
    3 if 72 hours

FIGURE 10.2   Indwelling catheter prep design evaluated at the subclavian, femoral, and median cubital sites.

data to assess the immediate and persistent antimicrobial effects of the two test products, the vehicle, and the reference product. A pooled, 2-factor analysis of variance (ANOVA) model should be used. Statistical summaries will be generated in both tabular and graphical form for the three test sample times for the five selected anatomical sites.

The anatomical sites evaluated will be handled separately in this evaluation. The level of significance for type I or α error is set at 0.05. However, all statistical comparisons will also be given as p values of observing a t calculated value as extreme or more extreme than that value, given the null hypothesis is true.

E. SAMPLE PROTOCOL: FULL BODY PRESURGICAL WASH

1. Purpose

Before surgical procedures are performed, it is standard that the proposed operative site be prepared with an effective antimicrobial to reduce the microbial populations residing on the skin and, hence, the potential for surgery-associated infection. Povidone iodine and CHG have been the two antimicrobials most commonly selected for preoperative patient skin preparation through the years. In efficacy trials with human volunteers whose baseline counts exceeded 10^5 organisms/cm^2, both antimicrobial products usually demonstrated at least a 3.00 $\log_{10}$ reduction in resident skin flora within 10 minutes of exposure.

Because the normal resident microbial populations of the abdominal and thoracic regions average approximately 10^3 organisms/cm^2, the vast majority of these flora are removed during the skin preparation process. At such anatomic sites as the inguinal region, however, where the microbial counts average 5.5×10^5 organisms/cm^2, the preoperative patient preparation, alone, may not be adequate in preventing postsurgical infection. In these anatomic areas with particularly high microbial counts, there will likely be significant numbers of microorganisms remaining on the prepared skin site. These organisms may have potential to cause postoperative infection, especially in immunocompromised patients.

Healthy human subjects older than 18 years, but younger than 70 years, are recruited. Insofar as possible, subjects are of mixed age, sex, and ethnic background. All subjects are free of clinically evident dermatoses and injuries to the skin areas being sampled. No subject will be admitted to the study who is using topical or systemic antimicrobial agents, or any other medication known to affect the normal microbial populations of the skin.

Once the subjects are admitted to the study, a 14-day pretest period is observed. During this time, subjects avoid the use of medicated soaps, lotions, shampoos, deodorants, chlorinated water baths, and ultraviolet light tanning beds, as well as skin contact with solvents, acids, and bases. This regimen permits stabilization of the normal microbial flora populations on the skin.

A 7-day baseline period follows the pretest period. Baseline skin sampling is conducted on days 1, 3, and 5 at both the abdominal and inguinal regions. A 5-day test period follows. Subjects use the test products, per instructions.

On test day 1, immediately after the shower procedure using the test products, each subject dries his or her body with a supplied, sterile, soft, absorbent terry towel.

There is a very good product that can be applied and has been approved as a bathing product. Each is then sampled immediately (within 10 minutes of showering) using the cylinder sampling technique and again at 3 and 6 hours after the shower wash. Identical sampling procedures are repeated on test days 2 and 5. Abdominal samples are taken from the region extending approximately 1 inch to the left and right of the umbilicus. The inguinal region is sampled at the uppermost inner aspect of the thigh in the inguinal crease. A sterile gauze material secured with adhesive tape is used to protect the sampling sites from transient microorganism contamination between samplings. The gauze bandage is removed for the 3- and 6-hour skin samplings and then reapplied. The subjects are not permitted to take additional showers or baths during the test period.

a.   Cylinder Sampling Technique

The cylinder sampling technique will be performed as follows: At the designated sampling times, a sterile cylinder will be held firmly onto the skin surface of the test site to be sampled, and 2.5 mL of sterile 0.1 M phosphate-buffered stripping solution with 0.1% Triton X-100 and product neutralizers will be instilled into the cylinder, and the skin area inside the cylinder will be massaged in a circumferential manner for 1 minute with a sterile rubber policeman. The entire procedure will be repeated without lifting the cylinder to obtain a second aliquot, and the two aliquots of 2.5 mL each will then be pooled.

One-milliliter (1-mL) aliquots of the fluid microorganism suspension (10^0 dilution) will be removed, plated, and/or serially diluted (as appropriate) with Butterfield's phosphate-buffered solution. Pour plates will be prepared in triplicate with 1 mL from each of these dilutions using soybean-casein digest agar, and the plates will be incubated at $30 \pm 2°C$ for 48 to 72 hours. Those dilutions yielding 25 to 250 colonies per plate will be preferentially counted. In cases where no plates provide counts in the 25 to 250 range, those plates with the highest number of countable colonies closest to that range or those from the lowest dilution will be used.

b.   Data Handling and Statistics

The estimated $\log_{10}$ number of viable microorganisms per square centimeter recovered from each sample site will be designated the R value. To convert the volumetric quantities collected during sampling into the number of colony-forming units per square centimeter (cm^2), the following formula will be employed:

$$R = \log_{10}\left[\frac{F\left(\dfrac{C_i}{n}\right)10^{-D}}{A}\right],$$

where:

R = the average colony count in $\log_{10}$ scale per cm^2 of sampling surface.

F = total number of mL of stripping fluid added to the sampling cylinder. In this study, F is equal to 5 mL.

$\left(\dfrac{C_i}{n}\right)$ = average of the triplicate plate counts used.

D = dilution factor of the plate counts.

A = inside area of the cylinder in cm^2.

Note: A $\log_{10}$ transformation will be performed on the collected data to convert them to a linear scale. A linear scale, more appropriately a $\log_{10}$ linear scale, is a basic requirement of the statistical models used in this study.

Statistical Analysis

The MiniTab (or other) statistical computer package will be used for all statistical calculations. An EDA, including stem-leaf ordering, letter value plots, and box plots, will be generated to assure that the R values approximate a normal (Gaussian) distribution [91]. A blocked, two-factor ANOVA test will be conducted and confidence intervals determined for the sample periods using the 0.05 level of significance for type I (α) error. Any outlier values obtained in the analysis will be noted. If laboratory error is determined to be the cause, these data will not be used in the statistical evaluation; however, they will be included in the raw data.

A pre-post experimental design will be utilized to evaluate and compare the antimicrobial effectiveness of the products.

Pre-Product Application	Post-Product Application
$R \; A \; O_{A,BL,1} \; O_{A,BL,2}$	$O_{A,T,10min} \; O_{A,T,30min} \; O_{A,T,6hr}$
$R \; A \; O_{A,BL,3} \; O_{A,BL,2}$	$O_{A,T,10min} \; O_{A,T,3hr} \; O_{A,T,6hr}$
$R \; C \; O_{C,BL,1} \; O_{C,BL,2}$	$O_{C,T,10min} \; O_{C,T,30min} \; O_{C,T,6hr}$
$R \; A \; O_{C,BL,3} \; O_{A,BL,2}$	$O_{A,T,10min} \; O_{A,T,3hr} \; O_{A,T,6hr}$

where:

R = Subjects randomly assigned to each of the two groups. A random number series will be generated for this purpose.

A = Independent variable: In this study, A represents the test product.

C = Independent variable: In this study, C represents the reference product.

O_{xij} = Dependent variable: For test or control product (A or C), microbial counts per cm^2 at the ith period and the jth time.

x = test (T) or control (C) product.

i = baseline (BL) or test (T) period.

j = Period 1, 2, or 3 for the baseline phase and times of 10 minutes, 3 hours, or 6 hours for the test phase.

References

1. D. S. Paulson, Quality Assurance of Topical Antimicrobials, *Pharmaceutical & Cosmetic Quality*, Nov/Dec, 26–32, 1997.
2. D. S. Paulson, Research Designs for the Soaps and Cosmetic Industry: A Basic Approach, *Soap/Cosmetics/Chemical Specialties*, Nov, 50–58, 1995.
3. D. M. Newman. *Sociology*. Sage, Thousand Oaks, CA, 1997.
4. H. R. Moskowitz. *Cosmetic Product Testing: A Modern Psychospiritual Approach*, Vol. III, Marcel Dekker, Inc., New York, 1984.
5. K. Wilber. *The Eye of Spirit*. Shambhala, Boston, 1997.
6. K. Wilber. *Sex, Ecology, Spirituality*. Shambhala, Boston, 1995.
7. R. Searle. *The Construction of Social Reality*. Free Press, New York, 1995.
8. D. O. Sears, L. A. Peplau, S. E. Taylor. *Social Psychology, 7th ed.* McGraw-Hill, Englewood Cliffs, NJ, 1991.
9. R. Kegan. *In Over Our Heads*. Harvard, Cambridge, 1994.
10. K. Wilber. *A Brief History of Everything*. Shambhala, Boston, 1995.
11. W. C. Frazier & D. C. Westhoff. *Food Microbiology, 4th ed.* McGraw-Hill, New York, 1988.
12. D. S. Paulson. Foodborne Disease: Controlling the Problem, *Environmental Health*, May, 15–19, 1997.
13. D. S. Paulson. Designing a Handwash Efficacy Program, *Pharmaceutical & Cosmetic Quality*, Jan/Feb, 42–44, 1997.
14. D. S. Paulson. A Broad-Based Approach to Evaluating Topical Antimicrobial Products, *Handbook of Disinfectants and Antiseptics*, J. M. Ascenzi, Ed. Marcel Dekker, New York, 1996.
15. J. Neter & W. Wasserman. *Applied Linear Statistical Models*. Irwin, Homewood, IL, 1974.
16. D. S. Paulson. Comparative Evaluation of Five Surgical Scrub Formulations, *Association of Operating Room Nurses Journal*, 60, no. 2, 246–256, 1994.
17. T. Peters. *Liberation Management*. Knopf, New York, 1992.
18. D. S. Paulson. Designing a Handwash for Healthcare Workers, *Soap/Cosmetics/Chemical Specialties*, June, 1996.
19. D. S. Paulson. To Glove or To Wash: A Current Controversy, *Food Quality*, June/ July, 60–63, 1996.
20. D. S. Paulson. A Proposed Evaluation Method for Antimicrobial Hand Soaps, *Soap/Cosmetics/Chemical Specialties*, June, 64–67, 1996.
21. M. J. Marples. *The Ecology of Human Skin*. Charles Thomas, Springfield, IL, 1965.
22. C. R. Leeson & T. S. Leeson. *Histology, 3rd ed.* W. B. Saunders, Philadelphia, 1976.
23. W. Montagna, A. M. Kligman, & K. S. Carlisle. *Atlas of Normal Human Skin*. Springer-Verlag, New York, 1992.
24. F. N. Maarzulli & H. I. Maibach. *Dermatoxicology, 3rd ed.* Hemisphere Publishing, New York, 1987.
25. H. I. Maibach & R. Aly. *Skin Microbiology*. Springer-Verlag, New York, 1981.
26. H. Schaefer & T. E. Redelmeier. *Skin Barrier*. Karger, Basel, Switzerland, 1996.
27. H. Mukhtar. *Pharmacology of the Skin*. CRC Press, Boca Raton, FL, 1992.
28. M. Schaechter, G. Medoff, & B. I. Eisenstein. *Mechanisms of microbial disease, 2nd ed.* Williams & Williams, Baltimore, 1993.
29. C. A. Mims. *The Pathogenesis of Infectious Disease, 3rd ed.* Academic Press, New York, 1987.

30. D. S. Paulson. Developing Effective Topical Antimicrobials, *Soaps/Cosmetics/ Chemical Specialties*, Dec, 50–58, 1997.

31. D. G. Maki. Infections Caused by Intravascular Devices Used for Infusion Therapy: Pathogenesis, Preventions, and Management, In: *Infections Associated with Indwelling Medical Devices*, A. L. Bisno & F. A. Waldvogal, Eds., American Society for Microbiology Press, Washington, D.C., 155–212, 1994.

32. W. K. Joklik, H. P. Willet, D. B. Amos, & C. M. Wilfert. *Zinsser Microbiology, 20th ed.* Appleton & Lange, Norwalk, CN, 387–400, 1992.

33. D. H. Groeschel & T. L., Pruett. Surgical Antiseptics. In: *Disinfection, Sterilization, and Preservation, 4th ed.* S. S. Block, ed., Lea & Febiger, Philadelphia, 1991.

34. D. S. Paulson. Designing a Healthcare Personnel Handwash, *Soap/ Cosmetics/ Chemical Specialties*, August, 53–57, 1988.

35. L. E. Hood, I. L. Weissman, W. B. Wood, & J. H. Wilson. *Immunology, 2nd ed.* Benjamin/ Cummings, Menlo Park, 1984.

36. W. L. Weissman, L. E. Hood, & W. B. Wood. *Essential Concepts in Immunology.* Benjamin/Cummings, Menlo Park, 1978.

37. J. Klein. *Immunology: The Science of Self-Nonself Discrimination.* John Wiley, New York, 1982.

38. J. Kuba. *Immunology, 2nd e*d. W. H. Freeman, San Francisco, 1991.

39. M. S. Thaler, R. D. Klausner, & H. J. Cohen. *Medical Immunology*, J. B. Lippin-cott, Philadelphia, 1977.

40. B. Benacerraf & E. R. Unanue. *Textbook of Immunology.* Williams & Williams, Phila-delphia, 1979.

41. E. J. Baron, L. R. Peterson, & S. M. Feinegold. *Bailey and Scott's Diagnostic Microbiology, 9th ed.* Mosby, St. Louis, 1994.

42. T. D. Brock, D. W. Smith, & M. T. Madigan. *Biology of Microorganisms, 4th ed.* Prentice-Hall, Englewood Cliffs, NJ, 1984.

43. R. M. Atlas. *Handbook of Microbiological Media*, CRC Press, Boca Raton, 1993.

44. B. A. Freeman. *Burrows Textbook of Microbiology, 22nd ed.* W. E. Saunders, Phila-delphia, 1985.

45. R. Y. Stanier, E. A. Adelberg, & J. Ingraham. *The Microbial World, 4th ed.* Prentice-Hall, Englewood Cliffs, NJ, 1976.

46. M. J. Pelezar, R.'D. Reid, & E. C. S. Chan. *Microbiology.* McGraw-Hill. 1977.

47. P. R. Murray, E. J. Baron, M. A. P. Fallen, F. C. Tenover, & R. H. Yolken. *Manual of Clinical Microbiology, 6th ed.* ASM Press, Washington, D. C., 1995.

48. P. L. Carpenter. *Microbiology, 4th ed.* W. B. Saunders, Philadelphia, 1977.

49. B. D. Davis, R. Dulbecco, H. N. Eisen, & H. S. Ginsberg. *Microbiology, 3rd ed.* Harper & Row, Cambridge, 1980.

50. J. D. Watson, N. H. Hopkins, J. W. Roberts, J. A. Steitz, & A. M. Weiner. *Molecular Biology of the Gene, Vol 1*, 4th ed. Benjamin/Cummins, Menlo Park, 1987.

51. J. F. MacFadden. *Biochemical Tests for the Identification of Medical Bacteria.* Williams & Williams, Baltimore, 1980.

52. G. M. Cooper. *The Cell: A Molecular Approach.* ASM Press, Washington, D.C., 1997.

53. L. Stryer. *Biochemistry, 3rd ed.* W. H. Freeman, San Francisco, 1990.

54. A. D. Russell. *The Destruction of Bacterial Spores.* Academic Press, New York, 1982.

55. C. Mims, A. Nash, J. Stephen, *Mim's Pathogenesis of Infectious Disease*, fifth ed. (San Diego: Academic Press, 2001).

56. M. G. Darby, G. A. O'Toole, "Microbial biofilms: From ecology to molecular genetics," *Microbiology and Molecular Biology Reviews 64* (December 2000) 847–867.

57. A. L. Reysenbach, E. Shock, "Merging genomes with geochemistry in hydrothermic ecosystems," *Science* 296 (May 10, 2002) 1077–1082.

58. J. Jass, S. Sunnan, J. T. Waller, "Microbial biofilms in medicine," in *Medical Biofilms: Detection, Prevention and Control*, eds., H. Jass, S. Surman, J. Waller (West Sussex, UK: John Wiley & Sons, Inc, 2003) 1–28.

59. M. Hentzer, M. Givskov, L. Eberl, "Quorum sensing in biofilms: Gossip in slime city," in *Microbial Biofilms*, eds., M. Ghannoum, G. A. O'Toole (Washington, DC: American Society of Microbiology, 2004) 118–140.

60. S. N. Wai, Y. Mizunoe, J. Jass, "Biofilm-related infections on tissue surfaces," in *Medical Biofilms: Detection, Prevention and Control*, eds., J. Jass, S. Sunnan, J. Waller (West Sussex, UK: John Wiley & Sons, Inc, 2003) 1–28.

61. F. Gotz, G. Peterson, "Colonization of medical devices by coagulase-negative Staphylococci," in *Infections Associated with Indwelling Medical Devices*, third ed., F. A. Waldvogel, A. L. Bisno, eds. (Washington, DC: American Society of Microbiology, 2000) 55–88.

62. J. M. Anderson, R. E. Marchant, "Biomaterials: Factors favoring colonization and infection," in *Infections Associated with Indwelling Medical Devices*, third ed., F. A. Waldvogel, A. L. Bisno, eds. (Washington, DC: American Society of Microbiology, 2000) 89–109.

63. S. E. Cramton, F. Gotz, "Biofilm development in Staphylococcus," in *Microbial Biofilms*, eds., M. Ghannoum, G. A. O'Toole (Washington, DC: American Society of Microbiology, 2004) 64–84.

64. D. Spratt, "Dental plaque," in *Medical Biofilms: Detection, Prevention and Control*, eds. J. Jass, S. Sunnan, J. Waller (West Sussex, UK: John Wiley & Sons, Inc, 2003) 1–28.

65. P. E. Kolenbrander, R. J. Palmer, "Human oral bacterial biofilms," in *Medical Biofilms: Detection, Prevention and Control*, eds. J. Jass, S. Sunnan, J. Waller, (West Sussex, UK: John Wiley & Sons, Inc, 2004) 85–117.

66. J. G. Thomas, G. Ramage, J. L. Lopez-Ribot, "Biofilms and implant infections," in *Microbial Biofilms*, eds. M. Ghannoum, G. A. O'Toole (Washington, DC: American Society of Microbiology, 2004) 269–293.

67. J. M. Steckelburg, D. R. Osman, "Prosthetic joint infections," in *Infections Associated with Indwelling Medical Devices*, third ed., F. A. Waldvogel, A. L. Bisno, eds. (Washington, DC: American Society of Microbiology, 2000) 173–209.

68. A. Stein, M. Drancourt, D. Raoult, "Ambulatory management of infected orthopedic implants," in *Infections Associated with Indwelling Medical Devices*, third ed., F. A. Waldvogel, A. L. Bisno, eds. (Washington, DC: American Society of Microbiology 2000) 211–230.

69. N. Phillips, *Berry & Kahn's Operating Room Technique*, 10th ed. (St Louis: Mosby, 2000) 740–765.

70. G. D. Ehrlich, F. Z. Hu, J. C. Post, "Role for biofilms in infectious disease," in *Microbial Biofilms*, eds. M. Ghannoum, G. A. O'Toole (Washington, DC: American Society of Microbiology, 2004) 332–358.

71. P. S. Stewart, P. K. Mukherjee, M. A. Ghannoum, "Biofilm antimicrobial resistance," in *Microbial Biofilms*, eds. M. Ghannoum, G. A. O'Toole (Washington, DC: American Society of Microbiology, 2004) 250–268.

72. C. Walsh, Antibiotics: Actions, Origins, Resistance (Washington, DC: American Society of Microbiology, 2003).

73. J. Netting, "Sticky situations," *Science News Online* 160 (July 2001) 2. Also available at http://www.sciencenews.org/articles/20010714/bob12.asp (accessed 24 Jan 2005).

74. P. S. Stewart, "Multicellular resistance: Biofilms," *Trends in Microbiology* 9 (May 2001) 5, 204.

75. R. Bayston, "Biofilm infections on implant surfaces," in *Biofilms: Recent Advances in Their Study and Control*, first ed., L. V. Evans, ed. (Amsterdam: Harwood Academic Publishers, 2000) 117–131.

76. L. A. Mennel, "Prevention strategies for intra-vascular catheter-related infections," in *Infections Associated with Indwelling Medical Devices*, second ed., F. A. Waldvogel, A. L. Bisno, eds. (Washington, DC: American Society of Microbiology, 2000) 407–425.

77. M. R. Brunstedt et al, "Bacteria/blood/ material interactions. I. Injected and preseeded slime-forming Staphylococcus epidermis in flowing blood with biomaterials," *Journal of Biomedical Materials Research* 29 (April 1995) 455–466.

78. C. W. Emmons, C. H. Binford, J. P. Utz, & K. J. Kwon-Chung. *Medical Mycology, 3rd ed.* Lea & Febiger, Philadelphia, 1977.

79. G. S. Moore & D. M. Jaciow. *Mycology for the Clinical Laboratory*, Prentice-Hall, Reston, VA, 1979.

80. Y. Al-Doory. *Laboratory Medical Mychology*, Lea & Febiger, Philadelphia, 1980.

81. J. W. Wilson & O. A. Plunhalt. *The Fungous Diseases of Man*. University of California Press, Berkeley, 1965.

82. S. G. Dick. *Immunological Aspects of Infectious Diseases*, University Park Press, Baltimore, 1979.

83. Center for Disease Control and Prevention. Fungal Diseases: Histoplasmosis, CDC, February 13, 2014, http://www.cdc.gove/fungal/diseases/histoplasmosis/index.html (accessed April, 2012).

84. N. K. Denzin & Y. S. Lincoln. *Handbook of qualitative research*. Sage, Thousand Oaks, CA, 1994.

85. D. C. Montgomery. *Design and analysis of experiments, 4th ed.* John Wiley & Sons, New York, 1997.

86. W. J. Dixon & F. J. Massey. *Introduction to Statistical Analysis, 4th ed.* McGraw-Hill, New York, 1983.

87. D. S. Paulson. Statistical Evaluations: How They Can Aid in Developing Successful Cosmetics. *Soaps/Cosmetics/Chemical Specialties*, Oct. 1985.

88. R. E. Kirk. *Experimental Designs, 3rd ed.* Brooks/Cole, Pacific Grove, CA, 1995.

89. D. S. Paulson. Applied Statistical Designs for the Researcher. Marcel Dekker, Inc., New York, 2003.

90. G. W. Snedecor & W. C. Cochran. *Statistical Methods, 7th ed.* Iowa State Press, Ames, IO, 1980.

91. P. F. Velleman & D. C. Hoaglin. *Applications, basics, and computing of exploratory data analysis*. Duxbury Press, Boston, 1981.

92. J. D. Gibbons. *Nonparametric Methods for Quantitative Analysis*, Holt, Rinehart and Winston, New York, 1976.

93. W. J. Conover. *Practical Nonparametric Statistics, 2nd ed.* John Wiley & Sons, New York, 1980.

94. D. Polkinghorne. *Methodology for the Human Sciences*. State University of New York Press, Albany, 1983.

95. W. R. Borg & M. D. Gall. *Education Research, 5th ed.* Longman, White Plains, New York, 1989.

96. D. S. Paulson. Designing a Handwash Efficacy Program, *Pharmaceutical & Cosmetic Quality*, April, 17–20, 1996.

97. *Annual Book of Standards, Section H, Vol. 11.05*, ASTM West Conshohochen, PA, 1998.

98. D. S. Paulson. The Role of Quality Assurance in Effective Topical Antimicrobial Development, *Pharmaceutical & Cosmetic Quality*, Jan/Feb, 42–44, 1997.

99. R. H. Green. *Sampling Design and Statistical Methods for Experimental Biologists*. John Wiley & Sons, New York, 1979.

100. D. S. Paulson. Evaluation of Topical Antimicrobial Products. In: *Handbook of Disinfectants and Antiseptics*, J. M. Ascenzi, ed., Marcel Dekker, Inc., New York, 17–42, 1996.

101. S. Budavari. *The Merck Index, 12th ed.* Merck & Co., Whitehouse Station, NJ, 1996.

102. *The United States Pharmacopeia, 23rd ed.* The National Formulary (18th ed.). U. S. Pharmacopeial Convention, Inc., Rockville, MD, 1995.
103. W. Gottardi. Iodine and iodine compounds, *Disinfection, Sterilization, and Preservation, 4th ed.* S. S. Bloch, Ed. Lea & Febiger, Malvern, PA, 152–166, 1991.
104. W. G. Characklis & K. C. Marshall. *Biofilms*, John Wiley & Sons, New York, 1990.
105. R. C. Weast, Ed. *CRC Handbook of Chemistry and Physics, 4th ed.* CRC Press, Boca Raton, FL, 1984.
106. A. G. Gilman, L. S. Goodman, A. Gilman, eds. *The Pharmacological Basis of Therapeutics, 6th ed.* New York: Macmillan, 1980.
107. G. W. Denton, "Chlorhexidine," in *Disinfection, Sterilization and Preservation*, 5th ed., S. S. Block, ed. Philadelphia: Lippincott, Williams & Wilkins, 321–335, 2001.
108. S. F. Bloomfield. Chlorhexidine and iodine formulations, *Handbook of Disinfectants and Antiseptics*, J. M. Ascenzi, Ed. Marcel Dekker, Inc., New York, 133–158, 1996.
109. N. S. Ranganathan. Chlorhexidine, *Handbook of Disinfectants and Antiseptics*, J. M. Ascenzi, Ed. Marcel Dekker, Inc., New York, 235–264, 1996.
110. E. L. Larson & H. E. Morton. *Alcohols, Disinfection, Sterilization, and Preservation*, S. S. Bloch, Ed. Lea & Febiger, Malvern, PA, 191–203, 1991.
111. M. L. Roller. Alcohols for antisepsis of hands and skin, *Handbook of Disinfectants and Antiseptics*, J. M. Ascenzi, Ed. Marcel Dekker, Inc., New York, 177–233, 1996.
112. Food and Drug Administration. Tentative Final Monograph for Healthcare Antiseptic Drug Products: Proposal Rule. Federal Register 59 (116) pp 31402–31452. June 17, 1994.
113. W. W. Daniel. *Applied Nonparametric Statistics*. Houghton-Mifflin Co., Boston, 1978.
114. D. S. Paulson. "Marketing Opportunities for Topical Anti-Infective Products," *Soaps/Cosmetics/Chemical Specialties*, Oct., 31–34, 1994.

Index